From Understanding to Action

ALLIANCE FOR GLOBAL SUSTAINABILITY BOOKSERIES
SCIENCE AND TECHNOLOGY: TOOLS FOR SUSTAINABLE DEVELOPMENT

VOLUME 5

Aims and Scope of the Series

The aim of this series is to provide timely accounts by authoritative scholars of the results of cutting edge research into emerging barriers to sustainable development, and methodologies and tools to help governments, industry, and civil society overcome them. The work presented in the series will draw mainly on results of the research being carried out in the Alliance for Global Sustainability (AGS).
The level of presentation is for graduate students in natural, social and engineering sciences as well as policy and decision-makers around the world in government, industry and civil society.

From Understanding to Action

Sustainable Urban Development in Medium-Sized Cities in Africa and Latin America

Edited by

Marco Keiner
ETH Zürich,
Switzerland

Christopher Zegras
Massachussetts Institute of Technology,
Cambridge, MA, U.S.A.

Willy A. Schmid
ETH Zürich,
Switzerland

and

Diego Salmerón
ETH Zürich,
Switzerland

Springer

A C.I.P. Catalogue record for this book is available from the Library of Congress.

ISBN 1-4020-2879-2 (HB)
ISBN 1-4020-2921-7 (e-book)

Published by Springer,
P.O. Box 17, 3300 AA Dordrecht, The Netherlands.

Sold and distributed in North, Central and South America
by Springer,
101 Philip Drive, Norwell, MA 02061, U.S.A.

In all other countries, sold and distributed
by Springer,
P.O. Box 322, 3300 AH Dordrecht, The Netherlands.

Printed on acid-free paper

springeronline.com

Printed in the Netherlands.

ALLIANCE FOR GLOBAL SUSTAINABILITY

An International Partnership

Alliance for Global Sustainability International Advisory Board

TABLE OF CONTENTS

CHAPTER 2

FRAMEWORK FOR RESEARCH ON SUSTAINABLE URBAN DEVELOPMENT

The Innovative Collaboration and Research Methodology of the DIMSUD Project

Diego Salmerón, Marco Keiner, Willy A. Schmid

CHAPTER 3

THE CASE STUDY CITIES

Santiago de Chile, Johannesburg, Gaborone

Branko I. Cavric, Aloyse C. Mosha, Marco Keiner

CHAPTER 4

CHALLENGES FOR SUSTAINABLE URBAN DEVELOPMENT

Daniel K. Irurah, Björn Malbert, Pål Castell, Jaan-Henrik Kain, Branko I. Cavric, Aloyse C. Mosha

CHAPTER 5

OPPORTUNITIES AND STRATEGIES FOR SUSTAINABLE URBAN DEVELOPMENT

Daniel K. Irurah, Björn Malbert, Aly Karam, Pål Castell, Marco Keiner, Branko I. Cavric

ACKNOWLEDGEMENTS

The editors and authors would like to thank

- Nathalie Jacobs (Kluwer Academic Publishers)
- Dr. Joanne Kauffman (Alliance for Global Sustainability)
- Prof. Cristina Felsenhard, Andrea Carrión, Laura Blonda, Rosanna Forray, Prof. Sergio León and Karen Davis (Pontificia Universidad Católica de Chile)
- Prof. Michael Edén, Pontus Slätis (Chalmers University of Technology)
- Kerry Pile and Amina Ismail (Witwatersrand University)
- Prof. Dr. Ralph Gakenheimer, Prof. (em.) Tunney Lee, and Prof. John de Monchaux (MIT)
- Dr. Peter Marcotullio (United Nations University)
- Prof. Dr. Alan Gilbert (University College London)
- Arley Kim (ETH Zurich)
- Dr. Martina Koll-Schretzenmayr and Anita Schürch (ETH Zurich)
- Andreas Gähwiler and Oswald Roth (ETH Zurich)
- Monika Zimmermann (ICLEI)
- Markus Eggenberger (DEZA)
- Balthasar Thalmann, Tobias Oberli, Philip Dijkstra, Michael Gravenkamp, Christian Bührer, Sandra Demarmels, Aeneas Wanner (Students at the ETH Zurich)
- Adam Tell Metzger and Anjali Mahendra (Students at MIT) and the Students of the MIT Santiago Urban Planning Studio
- Dr. Jon-André Lys (Swiss Commission for Research Partnerships with Developing Countries; KFPE/DEZA)
- Representatives of the City and Metropolitan Councils, local planning authorities and national ministries in Gaborone, Johannesburg and Santiago de Chile

for their collaboration, feedback, criticism, support and patience.

PREFACE

Hans van Ginkel
Rector, United Nations University

The challenges of the world's future are linked to the growing share of the global population that will reside in urban areas. UN projections indicate that by 2030 the world's urban population share will rise to 60 percent. Of the two billion added to the global population, 99 percent will be added to the urban areas of the world. Of this number, 95 percent will be in countries of the developing world.

As most people will live in urban areas we had better work to build and organize them as both attractive and less resource consuming places. That is, to promote sustainable urban development is to promote the creation of dense human settlements that are livable and have reduced their impacts on larger scale ecosystems.

While much attention has been focused on the "mega-cities," those with a population of over 10 million, the amount of people living in these places will remain almost constant while the smaller and medium size cities will be the great absorbers of the world's urban population. Indeed, it is predicted that while the absolute number of people that will live in urban centers of 10 million or more will increase from approximately 263 to 375 million between 2000 and 2015, their share of the total urban population will only increase from 9.2 percent to 9.8 percent, a 6.34 percent increase. At the same time, those living in urban centers of 5 to 10 million will increase their share of the total urban population by 19.14 percent. Their numbers will increase from approximately 155 to 248 million. The largest share of people will be living in urban centers with less than 1 million.

This book provides us with a challenging and interesting effort to open up a new line of thought in urban development in the developing world. It focuses on the sustainable urban development of medium sized cities. Particularly it addresses, through case studies of three very different cities, how to create urban areas that contribute to sustainable environmental development.

The authors have made the deliberate choice to focus on urban areas of the 2nd echelon. That is, those places that are moving towards, but are not yet "mega" size and most probably will never be. This focus will improve our knowledge on how to prevent many of these centers from the environmental and social disasters that have befallen upon the current so-called "mega-cities" in all parts of the globe.

There are, at least, three important conclusions from this work. First, this book points to the difficulty of universalizing the concept of "sustainable urban development." What emerged was the need to provide a context for creating sustainable urban environments, shaped by the contours of the historic, economic, political and cultural context from which they developed. For example, an urban center's ability to provide services needed for high quality environments depends upon whether it has the means of collecting taxes and elaborating and implementing sustainability-oriented strategies. Certainly, the decentralization of fiscal authority must accompany that of political authority. Moreover, in order to realize the appropriate strategies, participation of the civic society is essential. It is also important that visions, strategies, legal rules and physical plans be harmonized (i.e. that they do not contradict each other). If not, development patterns can take place that hinder sustainable urban development. Amongst the cities around the world, the context for moving towards these goals varies and therefore their pathways toward sustainable urban futures will also vary.

The second important conclusion of this book is the need for monitoring changes through the use of appropriate and measurable indicators. In order to shape urban development indicators that are based upon the needs of the community are indispensable. These measurements feed into the tools and instruments for planning and decision-making that enables each city to find its own path to a more sustainable future, thus contributing to global sustainability.

Third, more than research, however, the *Designing, Implementing Measuring Sustainable Urban Development* (DIMSUD) project has offered an innovative and successful form of international academic research linked to practice. This effort has clearly shown that the implementation of the findings of DIMSUD and other related projects requires the will of national and municipal governments to put them into practice. Very often, the day-to-day work overcrowds theses authorities with work and also, the skills of the planning staff and necessary equipment (GIS) are not available. Thus, sustainable development will not be achieved unless additional awareness creation and practical training will be given. This book delivers invaluable information about those needs.

The rich details of the case studies bring out the unique challenges and opportunities for Johannesburg, Gaborone, and Santiago de Chile in their search for sustainable development pathways. The authors are to be commended for their efforts at both research and project implementation. They have provided an example of how to approach the difficult tasks that lay before us. This book will be of use to like-minded scholars and practitioners in other cities of the developing world.

Chapter 1

URBAN DEVELOPMENT IN SOUTHERN AFRICA AND LATIN AMERICA

The scope of the project "Designing, Implementing and Measuring Sustainable Urban Development" (DIMSUD)

Marco Keiner[1], Diego Salmerón[1], Willy A. Schmid[1], Iván Poduje[2]
[1]ETH Zurich, [2]Pontificia Universidad Católica de Chile

Abstract: This initial chapter gives a brief summary of the scope and goals of the DIMSUD project. The starting point is the urbanization process in the developing world. The world's population is growing tremendously. Particularly affected are developing countries in Africa and Latin America where strong migration fluxes lead to unprecedented urban growth. City sprawl engenders a series of environmental, ecological, and social problems for its inhabitants. Authorities are often unable to steer the development, most of all in so-called "mega cities." But are there also opportunities to help turn rapid urbanization into sustainable urban development? Cities that are rapidly growing and yet relatively small in comparison to mega cities have to act before their problems become unsolvable. Based on case studies in Johannesburg, Gaborone, and Santiago de Chile, the DIMSUD project analyzed the challenges and potentials for action in order to achieve more sustainable urban development.

Key words: Global urbanization, sustainable urban development, mega cities, rapidly growing medium-sized cities, project DIMSUD

1. INTRODUCTION

The overall goal of the DIMSUD project was to make a relevant contribution to the development of new solutions for enhancing sustainable urban development in medium-sized cities in Africa and Latin America by offering both a scientific forum and an urban laboratory for joint learning. DIMSUD aimed to make urban sustainability operational, meaning that workable responses were to be produced to meet the case study cities' challenges as well

M. Keiner et al. (eds.), From Understanding to Action, 1–24.

as give general recommendations to other cities striving for sustainable development.

However, it was not intended to develop a new definition of "urban sustainability." Instead of just varying existing definitions and approaches, for example, from McGranahan and Satterthwaite 2003, Perlman et al 1998, FARN (quoted in The Regional Environmental Center 2004) and the conferences URBAN21, Sustainable Cities (2000; see Section 6.5 of this chapter) and others, the project focused on the process of achieving sustainability rather than the goals.

The phenomenon and core challenges of global urbanization are shown in this introductory chapter. Chapter 1 also discusses the different understandings and the specific contexts of urban sustainability. It emphasizes medium-sized cities with governance problems. Chapter 2 illustrates the background, the research framework for the project and highlights the innovative methodology of research collaboration focused on practical problem solving by introducing a model of multi and transdisciplinary international academic collaboration. The case study cities of Gaborone, Johannesburg and Santiago de Chile are presented in Chapter 3, while Chapter 4 highlights the challenges for sustainable urban development in these cities. Opportunities and strategies for sustainable urban development, especially for participation, health care and environmental and resource management as well as backcasting approaches for visions of future development are discussed in Chapter 5, followed by a closer look on how sustainability indicators can be used for steering urban development (Chapter 6). The concluding Chapter 7 gives a generalized summary of existing or realizable strategies and recommends further measures, such as monitoring, controlling and good governance. The strategies and tools, as proposed in Chapters 5 to 7, if applied, are meant to improve the cities' capacity to successfully manage urban growth by promoting social equity, preserving the environment, ensuring economic progress and, thus, improve the living conditions for city inhabitants. In addition, the strategies and tools provide other cites with the opportunity to use these new working examples in order to adapt and adopt appropriate approaches for their specific contexts. In other words, the project's outputs offer practical lessons for other cities on how to steer urban development towards a more sustainable path.

2. GLOBAL URBAN GROWTH

The world is currently undergoing an urban demographic revolution: Between 1950 and 2000, the world's population grew from 2.5 billion to 6 billion. The UN Population Division (2001) estimates that in 2050, between

8 billion and 13 billion people will live on Earth. Today, more than 50% of the world's population lives in urban areas. By 2030, this rate will presumably increase to 60%. In developing countries, only 40% of the population currently lives in urbanized areas but an expected population growth rate of 2.3% in developing countries by 2030 (UN 1999) will lead to an even sharper rise in the urbanization level. Whereas in 1990, 1.4 billion people lived in the urban areas of the Third World, this number will increase to 3.8 billion in 2030. This means (i) that 80% of global growth of the urban population will take place in the poor countries of the southern hemisphere (UN 2001), and (ii) from 2000 to 2030, the urban population in developing countries will grow by 60 million people a year, effectively doubling in the period from 2000 to 2030. From 2020 onward, the majority of people in developing countries will live in metropolitan areas. Currently, Africa is the continent with the lowest urban population level worldwide at 37% (UN 2001). Despite the threat of an HIV/AIDS pandemic (see Chapter 4, Section 6) Africa is simultaneously experiencing the fastest population growth rate and especially the fastest *urban* population growth rate. For example in 2020, Africa's sub-Saharan urban population will approach 440 million or 46% of its projected total of 952 million (Moor and Warah 2001). Thus, in 2025, more than 70% of the African population is expected to live in cities (Toepfer 2002). In contrast to Africa and most Asian countries, the urbanization rate in Latin American countries has grown earlier. In 1960, half of the population of Latin America was already urban and by 2000, 72% were urban dwellers, with as much as 85% in Argentina, Venezuela, and Chile (Coy 2002). The UN forecasts that by 2030, 84% of Latin Americans will live in cities (UN 2001).

This ongoing urbanization in Latin America is a result of falling mortality rates, rapid internal migration, economic development, and advances in technology. However, Gilbert (1998) points out that metropolitan growth has slowed markedly in recent years, mainly because of declining fertility rates.

Figure 1-1 illustrates the "explosive" urban growth experienced in developing countries.

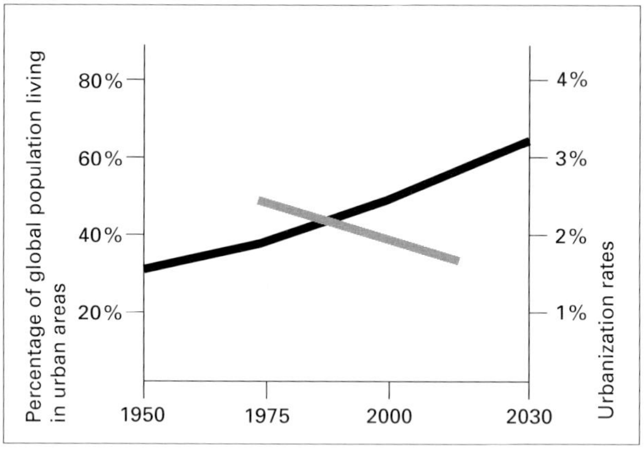

Figure 1-1. Development of global urban population 1950 to 2030 (Source: UN 1999)

Reducing the negative impacts of ongoing urbanization in developing countries appears to be one of the key challenges for the future of mankind (Brockerhoff 2000, Keiner 2004). Mitigating the effects of accelerated socioeconomic globalization and their threat to the natural environment is not only the responsibility of developing countries. In actuality, the destruction of the environment is mainly caused by activities in the urban regions of the highly developed world.

3. CITY SIZE AND GOVERNANCE

According to UN (1999), there are 28 cities worldwide that exceed the number of 8 million inhabitants or 19 cities with more than 10 million inhabitants each. These large cities are called "mega cities." Most of these mega cities are located in developing countries, mainly in Asia. Africa's mega cities are Lagos and Cairo. São Paulo, Mexico City, Buenos Aires, and Rio de Janeiro represent Latin America's mega cities. The UN (1999) expects that by 2015, there will be 325 cities with more than one million inhabitants and 27 "10+-million" cities worldwide, 23 of them in developing countries.

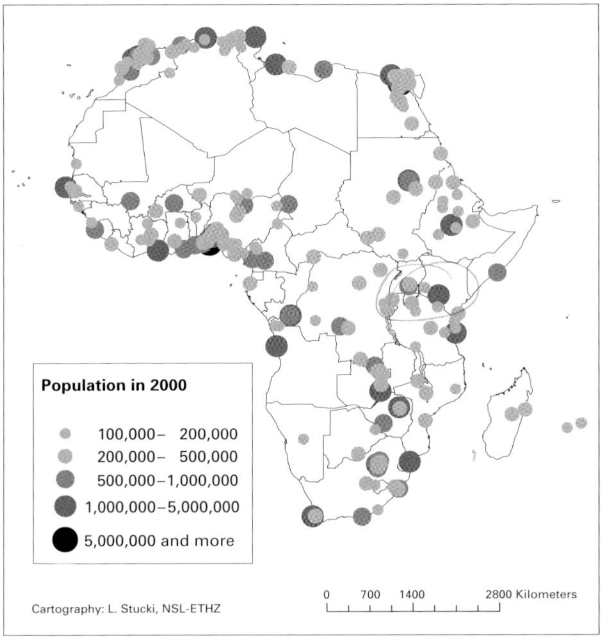

Figure 1-2. Africa's large cities (Data: UN 2001)

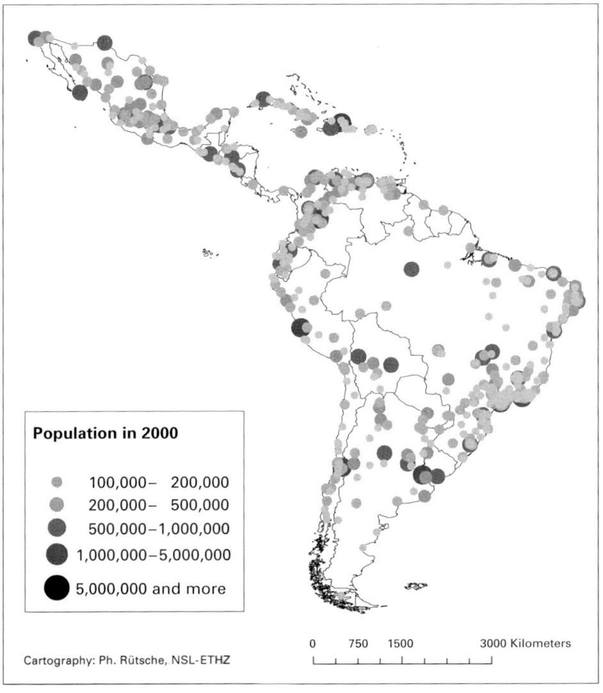

Figure 1-3. Latin America's large cities (Data: UN 2001)

In the last decade, a lot of research has been done on mega cities (see, for example, Perlman 1993, Fuchs et al. 1994, Gilbert 1996, Rakodi 1997, Bugliarello 1999, Fu-Chen and Marcotullio 2001). The authors have analyzed the driving forces for urbanization (demography, labor force change, etc), and their current and future impact on the city's ability to achieve sustainable development. The environmental and socioeconomic problems that need to be solved are central to these studies.

Two facts should be mentioned. First, grouping cities by their *size* alone (for example as "mega cities") does not make a lot of sense: Cairo, Tokyo, and Mexico City, for example, are mega cities. But what do they have in common? These cities vary substantially in every respect. What is more interesting than the cities' size is how they deal with the issues associated with it. Thus, a second issue is *governance capacity*. There are large cities that have been run well, have overcome environmental challenges and are increasingly clean, as there are smaller cities that are poorly managed and have terrible living conditions.

> *"A small city poorly managed is bad, a large city poorly managed may be as good (or as bad) as a small city well managed, but a large city well managed is definitely best."* (Prud'homme 1996)

Indeed, some have argued that neither the size of the city nor the speed of growth is most important in producing poor urban environments, but rather the main culprit is a lack of good governance (Satterthwaite 1996, Gilbert 1998).

Generally, large cities are more difficult to manage than smaller cities. Due to their huge size and the multitude and complexity of associated problems, achieving sustainable development in the mega cities of the developing part of the world seems to be a Sisyphean task. However, there are several examples of innovative projects leading to improved sustainability in big cities (Satterthwaite 1996, UN-HABITAT 2002). The prerequisite for their success has been good governance, i.e. the participation of individuals, households, communities, voluntary associations, and NGOs.

If not only the size of a city but also the city's governance capability are important to achieve sustainable urban development, the focus of sustainable development should not only be on mega cities but also on smaller cities with less successful governance. Data from the United Nations (UN 1999) show that in 1995 only 10.8% of the global urban population lived in cities of 8 million or more. This proportion will rise to 12.2% in 2015. The global share of the urban population living in cities of 5 to 8 million inhabitants in 1995 was 3.4% and will rise to 3.9% in 2015. On the other hand, those living in cities of between 1 and 5 million accounted for 23.5% of the global urban population in 1995, although this proportion is projected to decrease to

22.4% by 2015. The rest of us live and will continue to live in cities smaller than 1 million (over 60% of the urban population for both 1995 and 2015).

Although mega cities generally have a concentration of major universities and the best skilled technocrats as well as the most money, brainpower, and research attention, the much larger number of cities with less than 5 million inhabitants should not be ignored (Keiner and Schmid 2003).

4. CHALLENGES FROM URBANIZATION IN DEVELOPING COUNTRIES

For Africa, where fertility levels remain high and economies are weak, Brockerhoff (2000) stipulates that from 1960 to 1990, 75% of urban growth had been fueled by a natural population increase, compared with 51% in Asia. However, urbanization is not only characterized by natural population growth, but also by rural-urban migration (McGee 1998) and subsequent urban expansion. Big cities attract people from rural areas who hope to find work and better living conditions. In Latin America, for example, young women migrate to cities trying to diversify the overall family income by working in factories or in domestic and other services (Prothero and Chapman 1985, Bilsborrow 1996). In general, they stay in the cities, which leads to an impact on the gender-ratio both in rural and urban areas. However, migration is not necessarily a one-way process. In the case of Africa, there is a temporary or circular migration of mostly young men who try to find paid work after the harvesting season. After some months, they go back to their villages to till their fields. Such seasonal labor migration between urban and rural areas can lead to large swings in the population sizes of cities and the countryside.

Permanent migration can be observed through people who try to escape social control, the impacts of natural hazards like flooding or drought, and man-made disasters – most commonly in Africa – such as desertification, tribal conflicts, war, and social destabilization. For example, since 1970, more than 30 wars have been fought in Africa, the vast majority of them intra-State in origin. These armed conflicts account for more than 8 million refugees (Andam 2004).

Especially young people are pulled into the cities with the attraction of a modern lifestyle and better educational opportunities. However, this gain in individual freedom is linked to an erosion of traditionally strong social relationships and to a loss of social cohesion in the case of need (Chidester, Dexter and James 2003), even if kinship networks sometimes help to accommodate the new immigrants (Chabal and Daloz 1999). Moreover, mainly in Africa, rural to urban-bound migration leads to a loss of workforce in rural agriculture, entailing lower overall income in the rural areas, and if not compensated by higher grades of mechanization and use of fertilizers, a net reduction in agricultural productivity and possibly food scarcity.

In-migration leads to increased urban density, often combined with degradation of existing slum or "favela" quarters or the creation of new illegal settlements ("self-help housing") at the urban fringe. In some cases, as in Johannesburg after the end of the apartheid regime, even the city centers had been transformed into downgraded places where houses were illegally occupied and insecurity and crime were commonplace.

Indeed, if not properly managed, urban growth may become the source of a broad variety of problems (Liddle and Moavenzadeh 2002):

- Poor standards or even progressive decay in basic public infrastructure (energy supply, access to safe water, sewerage, schools, roads, preventive and curative health treatment, etc; see Chapter 4, Sections 6 and 7)
- Malnutrition, food scarcity, and diseases (HIV/AIDS, malaria, etc)
- Loss of fertile urban agricultural land for settlement purposes
- Lack of sanitation, which leads to groundwater pollution by nitrates and bacteria and causes infections from cholera to tuberculosis
- Poor drainage and poor waste management, entailing the contamination of rivers and streams by sewage outflows and waste disposal, which in turn leads to fresh water scarcity and consequently diseases
- Deforestation of the city's surroundings because of the need for fuel (of mostly low-income households), leading to an entire depletion of the vegetation and the start of gully erosion
- Air pollution and increased greenhouse gas emissions because of increasing traffic, industrial activity, and firewood and litter burning mixed with dust (see Chapter 4, Section 8.2 and Chapter 5, Section 2.1)
- Pressure on land use, poor land tenure security, and lack of affordable housing leading to unplanned settlements in the urban periphery and the mushrooming of nearby urban villages
- Poor urban design, neglected public parks and greenways (León 1998) as well as pressure on urban agriculture (Mosha and Cavric 2001)
- Inequality and urban poverty in general, combined with unemployment and low educational levels (see Chapter 4, Section 3)

Beyond the ongoing process of *urban transition*, at least three other major urban trends can be observed worldwide. The first is *urban deconcentration*, or the tendency of urban areas to expand outwards, most typically at rates faster than the overall population growth (producing net declines in overall densities of metropolitan built-up areas; see WBCSD 2001). The second trend that apparently accompanies urbanization and general economic growth is the increasing economic importance of *service industries*, which comprise a major share of economic activities in cities of industrialized countries and increasingly those in the developing world (Hall and Pfeiffer 2001). Finally, the continuously growing awareness of *urban environmental problems*, for example air pollution from increasing motor vehicle use, poor water quality, and their links with economic and social problems and challenges – now often referred to with the term "urban sustainability" – is a major factor influencing urban growth trends, interventions, and planning policies.

McGranahan et al. (1999 and 2001) detail the *Urban Environmental Transition Theory* (UET) and the environmental challenges faced by cities throughout development, but largely those of the developing world. The UET model suggests that as cities develop they are confronted with different sets of environmental problems. These challenges change in both scale (geographic and temporal) as well as type. Cities of the lowest incomes face the "brown" local issues of water supply, sanitation, waste removal, and indoor air pollution as most dominant. Rapidly developing cities face the broader regional problems related to chemical, industrial, and auto pollution. Finally, high-income cities face the global scale challenges of green house gas emissions, ozone depletion, and consumption related to changes in lifestyle. These impacts affect ecosystems and threaten the stability of the planet itself.

The problems listed above are among those well known to almost all large and rapidly growing medium-sized cities in the developing world. They lead directly to subsequent problems often associated with strategies for survival such as crime (drug use, robberies, sexual abuse), as well as the inner-urban segregation between rich and poor, and so forth. Inter-urban migration leading to segregation in Santiago de Chile will be highlighted in Chapter 4, Section 2.3.

Urban poverty in the developing world is already at a high level, and still it continues to increase. Cairncross et al. (1990) estimated that 600 million urban residents, 42% of the urban population, live in life- and health-threatening homes or neighborhoods. By 2025, half of the poor in Africa and even two-thirds of the poor in Latin America will be urban.

"Poverty" means more than only lack of income. It is also a lack of accessibility to public health services and other basic infrastructure; it is the exposure to environmental and social problems without any alternatives (McGranahan and Murray 2003).

Another danger for sprawling and densely populated cities in developing countries is their vulnerability to natural disasters, such as floods, hurricanes, mudslides, and earthquakes (Mitchell 1999, Pelling 2003). The more concentrated people are at any one place, the higher the risk of large losses of human life in the case of natural and also man-made disasters, for example, accidents in chemical industry and terrorism. Regarding ongoing global climate change and rising sea levels, flooding catastrophes such as those experienced lately in Bangladesh will also have to be taken into consideration; Cairo, Lagos, and most of all, a huge number of Asian cities are also located on coastal areas (World Bank 2003).

The fact is, not only do the aforementioned problems exist, but also it has not been possible to hinder them. Moreover, the intensity and dimension of these problems overwhelm the ability and capacity of local authorities to steer urban development into a desirable direction. Few time and money is available to resolve the problems, and as a result, there is few chance to come to workable solutions at all.

In many Latin American cities, urbanization problems appear to be unsolvable. The insufficient ability to manage urban growth was accompanied by the 1980's debt crisis, economic recession, structural adjustment policies, and hyperinflation – factors together leading to unemployment, low national investment, and declining living standards (Gilbert 1996).

In Africa, the boom of urbanization started only since the time of decolonization when new capital cities were founded. Unlike in Asia and Latin America, where large-scale investment created an impressive volume of new jobs in the manufacturing sector (for example, the "maquila" plants in Mexico), only a relatively small number of urban employment opportunities were created for city-bound migrants (Rogerson 1997). Today, black Africa's part in global trade as well as in industrial production is negligible (Le Monde Diplomatique 2003), its access to global markets is restricted, and its terms of trade are worsening (UNCTAD 2003). Net income from trade is very often used to service the debt payment, constraining investment and development. Thus, due to the lack of financial resources and foreign investment, the future freedom of action for Africa's national and urban governments will be very limited. Also, in many African states, there is the problem that political victory assumes a "winner takes all" form (Andam 2004), leading to changes in distribution and a lack of institutional sustainability. Moreover, African planning authorities very often suffer from mismanagement, because they are involved in corruption, lack efficiency and know-how, and tend to react instead of being proactive (Rakodi 1997, Chabal and Daloz 1999). Simultaneously confronted with the world's highest urban population growth rates, low economic growth, and political instability, Africa's mushrooming cities will face unprecedented challenges.

5. URBANIZATION AS DEVELOPMENT FACTOR

If we admit the historical comparison to Europe and North America, cities during the Industrial Era were motors of economic development, innovation, and interaction (Jacobs 1969 and 1984, Perlman et al. 1998). Urbanization in developing countries today may also lead to local clusters of entrepreneurial enterprise, and in compact cities (de Roo and Miller 2000), the efficiency of infrastructure investments is increased.

Southworth (1995), Prud'homme (1996) and Heinrich (2001) summarize the most outstanding aspects of agglomeration economies that benefit from the proximity between producers, suppliers, consumers, and workers:

- A larger urban labor market allows for the sufficient availability of a relatively inexpensive workforce, enhances the division and diversification of labor with new job opportunities, and leads to new skill combinations.
- Firms have access to a relatively big urban market without long transportation needs. There are opportunities for specialization and for innovation as well as the ability to react to changes in consumer's demands and potentials for sharing common inputs (warehousing, power etc).

According to the World Bank in 1989, about 60% of the GDP of developing countries was created in cities, and 80% of future GDP growth is estimated to occur there (World Bank 1990). Other related benefits of urbanization that accrue directly to people include issues of education, interaction, and transfer of know-how (Glaeser 1998). Indeed, big cities continue to be celebrated centers of culture and "social advancement."

Also, better health care facilities in cities are expected to augment the quality of life. However, HIV infection is higher in urban than in rural areas because of prostitution and liberal sex practices (UN 1994). Also, even if urban dwellers have spatial access to health treatment services, they often cannot afford them. On the other hand, if preventive and curative health treatment can also be made accessible for the poor, life expectancies will rise. The resulting aging of society may, in turn, engender new economic challenges to social security and welfare systems.

The opportunities of urbanization outlined above offer several keys to a successful continuation of urban development. However, other prerequisites are necessary in order to achieve a kind of urban development that moves towards sustainability.

6. ACHIEVING SUSTAINABLE URBAN DEVELOPMENT IN RAPIDLY GROWING CITIES

6.1 Different Understandings of Sustainable Development

Due to its importance to global sustainability, the idea of urban sustainability is, of course, not new. For example, the *Rio 1992 World Summit* was quickly translated into *Local Agenda 21* initiatives around the world. The *HABITAT II* conference (Istanbul 1996) proclaimed the right to appropriate living space for all people, and the world conference *URBAN21* (Berlin 2000) showed "best practice" examples of the sustainable development of cities. Sustainable urban development was also treated as a crosscutting theme in the *JOHANNESBURG SUMMIT 2002*.

Discussing sustainable development is like discussing other imprecise terms like "democracy," "human rights," and so forth. Although the Brundtland Commission has a standard definition for sustainable development (WCED 1987; see Figure 1-4), this definition has been subject to several modifications and was reformulated according to different points of view. Thus, the many definitions of sustainability vary considerably. A broad range of authors have proposed various principles, and strategies for achieving sustainable urban development (for example, Perlman et al. 1998, Haughton and Hunter 1994, Mitlin and Satterthwaite 1996) based on the overall definition of sustainable development given by the Brundtland Commission report:

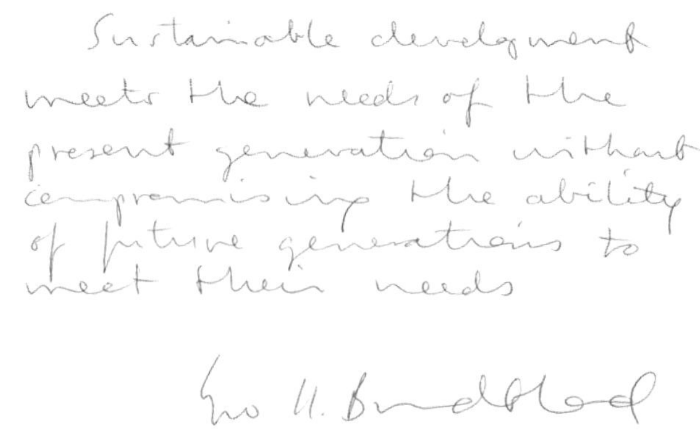

Figure 1-4. The original definition of "Sustainable Development" (autograph by Gro Harlem Brundtland)

Today, more than ever, disagreement exists as to the precise meaning of the terms *sustainability* and *sustainable development.* Bartlett (1997) examined different reports dealing with these terms. He concludes that the use of *sustainability* and *sustainable development* is often careless and contradictory. Thus, these terms are arbitrary and user-defined, and have lost their clear meaning. However, most definitions in use refer to the viability of natural resources and ecosystems over time, and to the maintenance of human living standards and economic growth. Depending on the specific historical, geographical, economic, and cultural context, the original and all subsequent definitions of sustainable development can be understood and interpreted in different ways.

Very broadly, we can group the many different interpretations into two ways of understanding sustainability: protection of the environment, and economic growth. The most obvious conflict since the World Summit 1992 in Rio is that developed countries mainly call for more respect of the environment and finite natural resources, whereas developing countries still claim economic development as the most important issue. The need to protect the environment and human living space is also recognized by the governments of poorer states, but not as a priority (see also Chapter 4, Section 8.1).

Consequently, the inherent conflict between (economic) development and maintaining the environment still exists in the conception of sustainability:

– Those who emphasize the first part of the definition proposed by the Brundtland Commission, that is, *"development that meets the needs of the present [...]"* remind us that our societies are far from responding in equal measure to the needs of all, be it on a world, regional or local scale. The question raised here predates the concept of sustainable development. It refers to the issue of equality among men, which dates from the 18th and 19th centuries, and emphasizes the foundations of the legitimacy of urban politics.
– Those who stress the second part, that is, *"[...] without compromising the ability of future generations to meet their own needs,"* focus on limiting the negative consequences that decisions in the present could have over the long run. This concern could translate into legislation, which seeks to ensure the "protection of the city" as a material entity, or into the evaluation of impacts on the environment.

In this context, sustainable development challenges economic development, raising the question of the *long run*. In some cases, the concern centers on the conservation of the material base of the natural and constructed space, while in other cases, the central issue is the *equity* of distribution of its benefits, giving sustainability a socio-political focus.

6.2 Urban Sustainability in a Regional and Global Context

The aim of the concept of sustainable development is to meet people's needs in a way that the environment does not suffer. Meeting people's needs depends on the use of resources, which are often non-renewable and generate waste. Both have a negative impact on the ability of the environment to continue its functions. The biggest share of resources is needed for urban centers, which produce pollution and waste having effects not only on the cities themselves but also beyond their boundaries, for example, greenhouse gas emissions. Thus, to solve the conflict between protecting the environment (inside and outside of cities) and meeting the people's economic needs, development has to become less consumptive and less waste creating while taking into account local, regional, national, and also global aspects. As McGranahan and Satterthwaite (2003) point out:

> *"The goal is not sustainable cities but cities that contribute to sustainable development within their boundaries, in the region around them, and globally."*

Thus, the governments of urban centers should be concerned about the impact of the city on the region and behave in a sustainable way even if the territories on which the city's waste and pollution have an effect (for example, forests as sinks for carbon dioxide emissions by city traffic and heating) lie outside the city government's jurisdiction (McGranahan and Satterthwaite 2003).

Perlman et al. (1998) reach a similar conclusion:

> *"There can be no sustainable city of the 21st Century without social justice and political participation, as well as economic vitality and ecological regeneration".*

During the preparatory meetings for the *URBAN21* Conference (Berlin, July 2000), the following vision of sustainable urban development was developed:

> *"Improving the quality of life in a city, including ecological, cultural, political, institutional, social and economic components without leaving a burden on the future generations. A burden, which is the result of a reduced natural capital and an excessive local debt. Our aim is that the flow principle, that is based on an equilibrium of material and energy and also financial input/output, plays a crucial role in all future decisions upon the development of urban areas."*

In order to emphasize on the pertinence and the applicability of the concept of sustainable development in the African and Latin American context, some regional specific definitions of "sustainable city" must be mentioned. One comes from the *Fundación Ambiente y Recursos Naturales* (FARN) in Argentina:

> *"A sustainable city integrates environmental dimensions into the social and economic sectors in order to meet the needs of current generations without compromising those of the future."* (FARN, quoted in The Regional Environmental Center 2004)

Also, the project "Sustainable Santiago" provides a definition, highlighting the need to put "environment" at the center:

> *"The project must guarantee that the municipal planning process includes effective methods oriented in the following themes that are a priority of sustainability:*

- *Environmental evaluation*
- *Improvement of air quality through urban design and management of transportation demands*
- *The reduction of industrial and municipal waste, as well as the conservation of water and energy."* (Proyecto Santiago Sustentable 1998)

6.3 The Need for Urban Sustainability

Regarding the challenges of global urbanization and the problems of large cities, the question that arises is not only how to achieve sustainable urban development, but how to avoid perpetuating negative and vicious cycles in it. It seems that we are far from what *Agenda 21* addresses in its Chapter 7, where the concept of promoting sustainable human settlements was introduced for the first time:

> *"Urbanization, if properly managed, offers unique opportunities for the supply of sustainable environmental infrastructure through adequate pricing policies, educational programs and access mechanisms that are economically and environmentally sound."*

Will a city ever be able to develop in a truly sustainable manner? According to the UN (1999), Lagos, for example, will nearly double its population from 2000 to 2015 to reach 23 million. This will exponentially increase the city's problems. Can the problems of Lagos be solved then? Of course, the earlier action is undertaken, the better the chances are of at least slowing down the increased quantity and complexity of urbanization problems. In other words, the smaller the city, the better the chances of curing its afflictions. So if the city decides to meet the challenges of urbanization, a lot of work has to be undertaken immediately: creating awareness among stakeholders, decision-makers and the public, defining locally specific objectives for sustainable urban development, elaborating measures in a participatory manner, budgeting financial resources for urban development (instead of for purchasing weapons, luxury merchandise from abroad, etc), and so on.

6.4 Meeting the Needs

One element, which remains to be discussed in the analysis of the concept of sustainability, is that of *necessity*. In all its aspects, development seeks to respond to individual and collective needs; needs which reflect different demands and will require different periods for resolution. In this area there are numerous debates, starting with the very definition of need. The stated needs must also be selected and prioritized, a process that implies focusing and excluding or postponing. A discussion about the order in which to focus and exclude must follow. The debate does not stop here, because the question of need is intimately tied to the issue of the direction of the change aimed at meeting these needs, that is, the direction of development.

According to Abers (1997 and 2001), Douglass and Friedmann (1997), Holston (1999), and Sandercock (2002), any project on urban sustainability must start by identifying the needs of the community. These needs must be incorporated into the design and implementation of sustainable policies.

Finally, before seeking optimum urban sustainability, in which the so-called four "E's" (*economic development, environmental protection, equity*, and *education*; some put *empowerment* and others put *ethics* or *aesthetics* instead of *education*) converge in a harmonious and proportional way, it will be necessary to understand several conflicts. The distortions which appear in the concrete implementation of policies, in the distribution of power and competences, in the face of pressures from economic groups and other stakeholders and in insufficient participation and networking between those who make rules all affect strategies and plans to implement the concept of sustainability.

6.5 Specific Contexts for Sustainable Urban Development

When analyzing the challenges (see Chapter 4) and opportunities (see Chapter 5) for sustainable urban development in the case study cities, one might be interested to ask *how* sustainable the cities' development is or could be. The "how" refers to measurable criteria. There are two ways to construct a framework in order to measure how well sustainability objectives have been achieved. First, one could use a general definition of sustainable urban development and describe the gap between the objectives and the actual development (distance-to-target analysis). Second, the same kind of analysis could be carried out on the basis of specific local definitions or visions of what sustainable urban development means in each case.

The 2000 Sustainable City Conference in Rio (quoted in ODPM 2004) stated:

> *"The concept of sustainability as applied to a city is the ability of the urban area and its region to continue to function at levels of quality of life desired by the community without restricting the options available to the present and future generations and causing adverse impacts inside and outside the urban boundary."*

Thus, there is an effort to integrate the diverse dimensions of social, economic and cultural development within the urban perimeter. However, the question centers on the manner in which equilibriums are generated between flows and stocks, in terms of the "urban metabolism" described by Acselrad in his matrix of discursive sustainability (Acselrad 1999). According to this author, these equilibriums are necessary in order for urban systems to attain the capacity to adapt to new requirements of development. Thus, the question arises as to the relevance of the "wise and scientific management of the territory" by public agents.

A closer reading of the above Sustainable City Conference definition reveals strong criticism of the tendency to comprehend the city as a closed system, unrelated to its regional, national and even global territories. The idea proposed instead is that cities are open systems, which require material and energy to keep their structures functioning. This perspective emphasizes the environmental dimension, in terms of ecological equilibriums or imbalances. However, there is no golden rule or single measure that can turn the development of cities in less developed regions into sustainable development. What measures have to be taken depend on the nature of the specific problems and the local and national context: culture, law, individual freedom, etc.

Working with a standard definition of sustainable urban development allows a comparison of the progress towards sustainability across several cities at a time. Some tools for comparison are the *Human Development Index* (HDI) and the *City Development Index* (CDI). The CDI proposes the use of benchmarks and ranks cities worldwide according to the criteria of Product (Income), Health, Education, Infrastructure, and Waste. However, this kind of benchmarking can be too general. On the other hand, if the possibility of comparison to other cities is limited, a more detailed and specific view makes it possible to distinguish different needs for sustainable development in each individual city.

In the case of Santiago, it is necessary to situate sustainable development within the framework of recent Chilean history. The Brundtland report appeared at the end of the period of Chile's military dictatorship (1973 to 1989), a period of considerable autocracy in political decision-making and in the determination of development goals. Chile's entrance in the process of globalization through a neo-liberal economic policy, the basic elements of which were to be maintained by the later democratic governments, was precocious within the Latin American context. The main challenges for sustainable urban development in Santiago are as follows (see Chapter 4):

- Air pollution: the city holds the dubious distinction, along with Mexico City and São Paulo, of suffering from the worst air pollution levels in Latin America (cf. Zegras and Litman 1997).
- Territorial socioeconomic segregation: In the municipal region of Santiago there are 34 socially homogeneous communities, clearly defined along rich-poor lines. There is, for example, an inter-urban migration of higher income groups towards new real estate projects on the periphery. Due to its role in generating economic activities and taxes for the local governments, the real estate market appears to be the main stimulant of the processes of urban development in Greater Santiago, especially those that spur expansive suburban growth.

Due to the apartheid legacy in South Africa, the city of Johannesburg has its own share of major socioeconomic and cultural challenges. Urban development in Johannesburg is still characterized by the following problems (see Chapter 4):

- Strong ethnic segregation: Even though initial mixing has started (for example, by Africans moving to the inner city and the leafy, well maintained northern suburbs, with better access to the city's amenities) most people have remained in highly segregated areas and townships. For instance, the densely populated and deteriorated areas of Soweto, Alexandra in the northeast and Orange Farm in the south still have close to 100% ethic African populations.
- Unemployment (now estimated at close to 40%) and a poor skills base, especially in view of a tertiary economy.
- Backlog in housing and services such as water, energy, sanitation, and waste disposal.
- Informal settlements (see Chapter 4, Section 5).

Gaborone, the capital of Botswana, is one of the fastest growing cities in sub-Saharan Africa, if not the fastest (Mosha 1996). It is the nation's focal center, where an overwhelming amount of both public and private investment is made. The main problems for urban development in Gaborone are as follows (see Chapter 4):

- Early master planning mistakes only take into account Gaborone's administrative function with a maximum of 20,000 inhabitants by the end of the 20-year planning horizon in 1983.
- Thus, the master plan of 1963 was characterized by a comparatively low-density form of development based on the Garden City model with a generous provision for pedestrian walkways, open spaces, and closely tied neighborhood units. Housing development was polarized with high and medium income on one side of the town and low income on the other; and the urban structure was such that it allowed little space for expansion outside of the original layout. In addition, the plan did not take into account any possible growth from migrating job seekers (Mosha 1996). In reality, the population in 1983 grew to threefold of the assumed figure. This underestimation of population growth brought with it a great deal of consequences, like shortages in serviced land for housing, as well as stress on the existing infrastructure and other facilities.
- The ongoing, uncontrolled, leapfrog expansion led to an overspill of Gaborone into the peri-urban settlements (Keiner and Salmerón 2003). Thus, the satellite settlements on tribal land around Gaborone have been growing at annual rates of 16% or more (Molebatsi 1996), becoming dormitory settlements.

In conclusion, the problems of the three case study cities differ, as do their size, culture, history, and geographical background. Each of these cities must find their own way toward sustainable urban development. However, there are some common challenges and tasks for urban management, which will be highlighted in Chapters 4, 5 and 7.

7. THE SCOPE: RAPIDLY GROWING MEDIUM-SIZED CITIES WITH GOVERNANCE PROBLEMS

In the past, research efforts and publications have placed significant focus on mega cities with 8 million inhabitants or more. Important is not only the size of a city, but also the speed of its change. In the future, more attention has to be focused on the problems of the fastest growing urban centers, almost all of which are located in the developing world. Hall and Pfeiffer (2000) use the term "cities of hyper growth" to describe cities in sub-Saharan Africa, India, the Islamic Middle East and poorer Latin America characterized by rapid population growth, an economy dependent on the informal sector, widespread poverty and sub-standard housing, basic environment and public health problems and governance problems. These problems of relatively small but rapidly growing cities are similar to those of mega cities, yet offer the opportunity for earlier intervention and correction.

Due to enormous growth rates, most of the sub-Sahara cities are on their way to becoming million-inhabitant or mega cities very soon. Due to the pace of their development, sustainable urban management is becoming more and more difficult, if not impossible. Not only mega cities have mega problems. As Hall and Pfeiffer (2000) suggest,

> *"Some of the biggest problems occur in relatively small cities."*

What counts is not the size alone, but the management ability to cope with problems. This is also true for Latin American cities between 2 and 5 million inhabitants like Santiago de Chile, Belo Horizonte, Porto Alegre, Recife, and many others (Gilbert 1996).

The UN (1999) stipulates that in 2015, the largest proportion of urban dwellers in the developing world will still live in cities less than 500,000 inhabitants. Montgomery et al. (2003) point out that very often, small cities (even those with less than 100,000 inhabitants) lack an adequate supply of fresh water, electricity, waste disposal infrastructure and schools. Thus, medium-sized and rapidly growing cities merit more scientific attention. The capitals of African countries, especially small ones like Gaborone with less than 250,000 residents (Mosha 1996), tend to become the sole economic and political center of the country. If they are twice as large as the second largest city in the country, these dominant cities are called primary cities. Haub (2001) points out the importance of this kind of city:

> *"While many countries simply do not have populations large enough to generate one or more mega cities, the influence of a primary city, usually the capital, can have just as significant an impact on national development as the urban giants."*

The UN (1999) estimates that in 2015, there will be about 564 cities worldwide with more than one million inhabitants. In 1975, there were only 195 such cities. The sharpest increase in new million-inhabitant cities will occur in the less developed regions of the world. Therefore early action is needed for rapidly growing cities to avoid the typical environmental, economic, and social problems million and mega cities are faced with.

Urban development planning towards sustainability is of top priority for rapidly growing cities in the developing world, even if they are sub-million cities (Brockerhoff 2000). These cities must orient their planning and development management towards the predictable future (El-Shakhs 1997). With this type of planning, future problems can be reduced, the management of metropolitan areas can be improved, and opportunities for future development towards sustainability can be prepared.

REFERENCES

Abers, R. 1997. Learning Democratic Practice: Distributing Government Resources through Popular Participation in Porto Alegre, Brazil. In M. Douglass and J. Friedmann (Eds.) *Cities for Citizens: Planning and the Rise of Civil Society in a Global Age*. Chichester (UK): John Wiley & Sons: 39–65.

Abers, R. 2001. Practicing Radical Democracy – Lessons from Brazil. *DISP* 147: 32–38.

Acselrad, H. 1999. Sustentabilidad y ciudad. *EURE Revista Latinoamericana de Estudios Urbanos y Regionales*, Vol. 25, No. 74.

Andam, K. 2004. *Main Challenges of Sustainable Urban Development in Fast Growing Cities of the South*. Keynote speech. Annual Meeting of the Alliance for Global Sustainability, Gothenburg, March 23, 2004.

Asian Development Bank (ADB) 2001. *Urban Indicators for Managing Cities*. Manila.

Bartlett, A.A. 1997. Reflections on Sustainability, Population Growth and the Environment. *Renewable Resources Journal*, Vol. 15, No. 4: 6–22.

Bilsborrow, R. (Ed.) 1996. *Migration, Urbanization, and Development: New Directions and Issues*. New York: United Nations Population Fund and Kluwer Academic Publishers.

Breheny, M. 1990. Strategic Planning and Urban Sustainability. *Proceedings of the TCPA Annual Conference, Planning for Sustainable Development*. London.

Brockerhoff, M.P. 2000. An Urbanizing World. *Population Bulletin*, Vol. 55, No. 3.

Bugliarello, G. 1999. Megacities and the Developing World. *The Bridge*, Vol. 29, No. 4.

Cairncross, S., J.E. Hardoy, and D. Satterthwaite (Eds.) 1990. *The Poor Die Young: Housing and Health in Third World Countries*. London: Earthscan Publishers.

Chabal, P. and J.P. Daloz 1999. *Africa Works – Disorder as Political Instrument*. Oxford: James Currey and Indiana University Press.

Chidester, D., P. Dexter, and W. James 2003. *What Holds us Together: Social Cohesion in South Africa*. Cape Town: HSRC Publishers.

Coy, M. 2002. Jüngere Tendenzen in der Verstädterung in Lateinamerika. In K. Bodemer, D. Nolte, and H. Sangmeister (Eds.) *Lateinamerika Jahrbuch 2002*. Frankfurt a.M.: Vervuert.

Devas, N. and C. Rakodi (Eds.) 1993. *Managing Fast Growing Cities: New Approaches to Urban Planning and Management in the Developing World*. Harlow, England: Longman.

Douglass, M. and J. Friedmann (Eds.) 1997. *Cities for Citizens: Planning and the Rise of Civil Society in a Global Age*. Chichester (UK): John Wiley & Sons.

El-Shakhs, S. 1997. Towards Appropriate Urban Development Policy in Emerging Mega Cities in Africa. In C. Rakodi (Ed.) *The Urban Challenge in Africa – Growth and Management of its Large Cities*. Tokyo, New York, Paris: UNU Press: 497–526.

Friedmann, J. 1986. The World City Hypothesis. *Development and Change*, 17: 69–83.

Friedmann, J. 2001. The Governance of City-Regions in East and Southeast Asia. *DISP* 145: 4–9.

Friedmann, J. 2002. *The Prospect of Cities*. Minneapolis, London: University of Minnesota Press.

Fuchs, R.J., E. Brennan et al. (Eds.) 1994. *Mega City Growth and the Future*. Tokyo, New York, Paris: UNU Press.

Gilbert, A. (Ed.) 1996. *The Mega-City in Latin America*. Tokyo, New York, Paris: UNU Press: 307–331.

Gilbert, A. 1998. World Cities and the Urban Future: The View from Latin America. In F.-C. Lo and Y.-M. Yeung (Eds.) *Globalization and the World of Large Cities*. Tokyo, New York, Paris: UNU Press: 174–202.

Gilbert, A. and J. Gugler 1991. *Cities, Poverty, and Development: Urbanization in the Third World*. New York: Oxford University Press.

Glaeser, E. 1998. Are Cities Dying? *The Journal of Economic Perspectives*, Vol. 12, No. 2: 139–160.

Hall, P. and U. Pfeiffer 2000. *Urban Future 21 – A Global Agenda for Twenty-First Century Cities*. London: Routledge.

Harrison, P. and F. Pearce 2001. *AAAS Atlas of Population and Environment*. New York: American Association for the Advancement of Science.

Haughton, G. and C. Hunter 1994. *Sustainable Cities*. Regional Policy and Development Series. Regional Studies Association. London and Bristol, Pennsylvania: Jessica Kingsley Publishers.

Heinrich, C. 2001. *ICT Infrastructure and Externalities Affecting Spatial Structures*. Working Paper of the Globalization and World Cities Study Group and Network. http://www.lboro.ac.uk/gawc/rb/rb67.html, retrieved on October 4, 2003.

Holston, J. 1999. *Cities and Citizenship*. Durham: Duke University Press.

Jacobs, J. 1969. *The Economy of Cities*. New York: Random House.

Jacobs, J. 1984: *Cities and the Wealth of Nations*. New York: Random House.

Keiner, M. 2004. Sustainable Development and Urban Management in Developing Countries. In Schretzenmayr, M., M. Keiner and G. Nussbaumer (Eds.) *The Real and the Virtual Worlds of Spatial Planning*. Heidelberg: Springer: 43–59.

Keiner, M. and D. Salmerón 2003. Sustainable Urban Development in Gaborone, Botswana. DIMSUD working paper. *ETH e-Collection*. http://e-collection.ethbib.ethz.ch/cgi-bin/show.pl?type=bericht&nr=308, retrieved on May 6, 2004.

Keiner, M. and W.A. Schmid 2003. Urbanisierungstendenzen in Entwicklungsländern – Probleme und Potenziale für nachhaltige Stadtentwicklung. *DISP* 155: 49–56.

Le Monde Diplomatique 2003. *Atlas der Globalisierung*. Berlin: taz Verlag.

Leon, S. 1998. Conceptos sobre Espacio Público, Gestión de Proyectos y Lógica Social. *EURE Revista Latinoamericana de Estudios Urbanos y Regionales*, No. 71.

Liddle, B. and F. Moavenzadeh 2002. Cities: Challenges and Opportunities for Sustainability. In F. Moavenzadeh, K. Hanaki and P. Baccini (Eds.) *Future Cities: Dynamics and Sustainability*. Alliance for Global Sustainability Book Series Science and Technology: Tools for Sustainable Development, Vol. 1. Dordrecht, Boston, London: Kluwer: 1–15.

Lo, F.-C. and P. Marcotullio (Eds.) 2001. *Globalization and the Sustainability of Cities in the Asia Pacific Region*. Tokyo, New York, Paris: UNU Press.

Lo, F.-C. and Y.-M.Yeung (Eds.) 1998. *Globalization and the World of Large Cities*. Tokyo, New York, Paris: UNU Press.

McGee, T.G. 1998. Globalization and Rural-Urban Relations in the Developing World. In F.-C Lo and Y.-M. Yeung (Eds.) *Globalization and the World of Large Cities*. Tokyo, New York, Paris. UNU Press: 471–496.

McGranahan, G., P. Jacobi, J. Songsore, C. Surjadii, and M. Kjellén 2001. *The Citizens at Risk: From Urban Sanitation to Sustainable Cities*. London: Earthscan.

McGranahan, G. and F. Murray 2003. *Air Pollution and Health in Rapidly Developing Countries*. Sterling VA: Stylus Publishing.

McGranahan, G. and D. Satterthwaite 2003. Urban Centers: An Assessment of Sustainability. *Annual Review of Environmental Resources,* Vol. 28: 243–274.

McGranahan, G., J. Songsore, M. Kjellén 1999. Sustainability, Poverty and Urban Environment Transitions. In D. Satterthwaite (Ed.) *The Earthscan Reader in Sustainable Cities*. London: Earthscan.

Mitchell, J.K. (Ed.) 1999. *Crucibles of Hazard: Mega-Cities and Disasters in Transition.* Tokyo, New York, Paris: United Nations University Press.

Mitlin, D. and D. Satterthwaite 1996. Sustainable Development and Cities. In C. Pugh (Ed.) *Sustainability, the Environment and Urbanization.* London: Earthscan.

Molebatsi, C.O. 1996. Towards a Sustainable City: Gaborone, Botswana. *Ambio* Vol. 25, No. 2. Stockholm: Royal Swedish Academy of Sciences: 126–133.

Moor, J.H. and R. Warah 2002. The State of the World's Cities 2001. In *Global Outlook. International Urban Research Monitor.* Washington: U.S. Department of Housing and Urban Development: 1–44.

Mosha, A.C. 1996. The City of Gaborone, Botswana – Planning and Management. *Ambio* Vol. 25, No. 2. Stockholm: Royal Swedish Academy of Sciences: 118–125.

Mosha, A.C. and B. Cavric 2001. Incorporating Urban Agriculture in Gaborone City Planning. *Urban Agriculture*, No. 4: 25–27.

Newman, P. and J. Kenworthy 1999. *Sustainability and Cities: Overcoming Automobile Dependence.* Washington, DC: Island Press.

Office of the Deputy Prime Minister (ODPM) 2004. *Assessing the Impacts of Spatial Interventions: Regeneration, Renewal and Regional Development – 'The 3Rs Guidance.'* http://www.odpm.gov.uk/stellent/groups/odpm_about/documents/page/odpm_about_029009-27.hcsp#P1867_385835, retrieved on June 22, 2004.

Pelling, M. 2003. *The Vulnerability of Cities – Natural Disasters and Social Resilience.* London: Earthscan.

Perlman, J. 1993 Mega-Cities: Global Urbanization and Innovation. *The Mega-Cities Project, Publication MCP-013.* Hartford.

Perlman, J., E. Hopkins and Å. Jonsson 1998. Urban Solutions at the Poverty/Environment Intersection. *The Mega-Cities Project, Publication MCP-018.* Hartford.

Prothero, R.M. and M. Chapman 1985: *Circulation in Third World Countries.* London: Routledge & Kegan Paul.

Proyecto Santiago Sustentable 1998. Sustainable Santiago. *Boletín* 2, February. http://cities21.com/iclei/boletin2.html, retrieved on October 3, 2003.

Prud'homme, R. 1996. Managing Megacities. *Le courrier du CNRS* No. 82: 174–176.

Rakodi, C. (Ed.) 1997. *The Urban Challenge in Africa.* Tokyo, New York, Paris: UNU Press.

Rakodi, C. 1998. Globalization Trends and Sub-Saharan African Cities. In F.-C. Lo, Y.-M. Yeung (Eds.) *Globalization and the World of Large Cities.* Tokyo, New York, Paris: UNU Press: 314–351.

Rogerson, C.M. 1997. Globalization or Informalization? African Urban Economies in the 1990s. In C. Rakodi (Ed.): *The Urban Challenge in Africa.* Tokyo, New York, Paris: UNU Press: 337–370.

Roo, G. de and D. Miller (Eds.) 2000. *Compact Cities and Sustainable Urban Development. A Critical Assessment of Policies and Plans from an International Perspective.* Aldershot: Ashgate.

Sandercock, L. 2002. Practicing Utopia: Sustaining Cities. *DISP* 148: 4–9.

Satterthwaite, D. 1996. Towards Healthy Cities. *People and the Planet* Vol. 5, No. 2.

Southworth, F.A. 1995. Technical Review of Urban Land Use – Transportation Models as Tools for Evaluating Vehicle Travel Reduction Strategies. In: *U.S. Federal Highway*

Administration Land Use Compendium. Prepared for the Travel Model Improvement Program. Washington, D.C.

Thale, T. 2002. *Jo'burg Plans to End Informal Settlements*. http://www.joburg.org.za/june_2002/housing.stm, retrieved on March 3, 2004.

The Regional Environmental Center 2004. *What is a Sustainable City? Defined by FARN, Argentina*. http://www.rec.org/REC/Programs/SustainableCities/FarnArgentina.html, retrieved on March 5, 2004.

Toepfer, K. 2002. *Zukunftsbeständige Stadt- und Regionalentwicklung – Leitmotiv für die Problembewältigung der Megacities*. Paper prepared for the International Conference "Megacities – Nightmare or Chance for Developing Countries." Hannover, June 17, 2002.

UN 1994. *AIDS and the Demography of Africa*. New York.

UN 1999. *World Urbanization Prospects – The 1999 Revision*. New York.

UN 2001. *The State of World Population*. http://www.unfpa.org, retrieved on October 3, 2003.

UNCTAD 2003. *Economic Development in Africa: Performances, Prospects and Policy Issues*. Geneva.

UN-HABITAT 2002. *Best Practices Database*. http://www.bestpractices.org, retrieved on October 10, 2003.

WCED The World Commission on Environment and Development 1987. *Our Common Future*. Oxford: Oxford University Press.

World Bank 1990. *World Development Report 1990: Poverty*. New York: World Bank and Oxford University Press.

World Bank 2003. *World Development Report 2003: Sustainable Development in a Dynamic World*. New York: World Bank and Oxford University Press.

World Business Council for Sustainable Development (WBCSD) 2001. *Mobility 2001: World Mobility at the End of the Twentieth Century and its Sustainability*. Prepared by the Massachusetts Institute of Technology and Charles River Associates for the WBCSD Sustainable Mobility Working Group.

Zegras, C. and T. Litman 1997. Cost Estimates of Transport Air Pollution in Santiago Chile. *Transportation Research Record No. 1587: Effects of Transportation on Energy and Air Quality*.

Chapter 2

FRAMEWORK FOR RESEARCH ON SUSTAINABLE URBAN DEVELOPMENT

The Innovative Collaboration and Research Methodology of the DIMSUD Project

Diego Salmerón, Marco Keiner, Willy A. Schmid
ETH Zurich

Abstract: This chapter discusses the importance of international academic collaboration on sustainable development and presents the innovative methodological framework for research collaboration within the DIMSUD project. Focus is laid upon four crosscutting research themes that enabled a comparative assessment of the three case study cities: Gaborone, Santiago de Chile, and Johannesburg. Finally, a discussion of research limitations will be followed by an outlook on forthcoming training needs and the dissemination of research results.

Key words: Background and goals, Alliance for Global Sustainability, methodological framework, transdisciplinary collaboration, research themes, urban design studio, dissemination

1. INTERNATIONAL ACADEMIC COLLABORATION ON SUSTAINABLE URBAN DEVELOPMENT AS BACKGROUND OF THE DIMSUD PROJECT

The fact that we live in a global village can no longer be overlooked (e.g. Shuman 1994, Labarca and Hendrickson 1999). Media such as television and the Internet are bringing information from all over the world to our eyes, increasingly often in real-time. Local concerns very often follow a more general pattern, leading to the awareness of global challenges, expressed locally. Steadily, the awareness that we live in one world and that everybody

M. Keiner et al. (eds.), From Understanding to Action, 25–39.

assumes responsibility for the Earth's future as a whole has been growing (e.g. Jonas 1985, Staub and Green 1992). This is not only due to the catalytic effect of the big World Conferences organized by the United Nations since the 1990's (e.g. *World Conference on Environment and Development* Rio 1992, *World Conference on Human Rights* Vienna 1993, *International Conference on Population and Development* Cairo 1994, *World Summit for Social Development* Copenhagen 1995, *World Conference on Women* Beijing 1995, *HABITAT II* Istanbul 1996, *World Food Summit* Rome 1996, *Rio+5* New York 1997, and so on) but also influenced by NGOs and the civic society, leading to ethical and moral guidelines like the *Earth Charter* (Keiner 2000) or grassroots-driven activities on the communal level (e.g. *Local Agenda 21*, encompassing intragenerational equity). Perhaps the most impelling evidence for the necessity to think and act with global responsibility has been gained by the "overview effect" or the perspective of astronauts and cosmonauts looking from outer space on our small and vulnerable blue planet with its thin atmosphere (White 1998).

In the past few years, scientific matters have also been increasingly taking on a global character. As pointed out in Chapter 1, urbanization is a global phenomenon and one of the key challenges for future human development (e.g. Moavenzadeh et al. 2002). The emerging need to address fundamental societal issues, such as sustainable development and the effects and impacts of global urbanization, requires the pooling of academic energies and know-how from "North" and "South" to solve problems of unsustainable urban development worldwide (e.g. Greene and Johnston 1997). Ultimately, globalization has contributed to a broader awareness of the opportunities that networking in sciences and the exchange of expertise offer for accelerating socioeconomic development and creating new strategies for recurring problems. In addition, globalization has led to greater international scientific cooperation between researchers aided by modern communications and computing (cf. UNESCO 1998).

International research collaboration is important for several reasons:

- Cooperation among scientists from different countries assembles intellectual capital for the production of new knowledge, benefiting from diverse cultural perspectives.
- Major research issues of mutual interest around the world, such as sustainable urban development demand consensus, and scientific collaboration, can be discussed in terms of common objectives, problems, and best practice recommendations.
- Scientific collaboration can promote development, awareness creation, and joint learning.
- Joint research programs can lead to improved communication between research and practice, putting innovation and knowledge to work.

The Alliance for Global Sustainability (AGS) is a worldwide partnership of university scholars and researchers (ETH Zurich, MIT, Chalmers, and University of Tokyo) who collaborate with business leaders, government policy makers, and environmentalists to identify and develop effective pathways to sustainable development. Its task is to enhance the study of large-scale, multidisciplinary environmental problems facing the world's ecosystems, economies, and societies (AGS 2003). For projects related to these objectives, the AGS grants seed money, thus offering invaluable opportunities for researchers to attack the most pressuring problems worldwide and come up with innovative and applicable solutions.

In 2000, ETH Zurich and the United Nations University signed a statement of intention aiming at establishing a research and training centre on urban development and sustainability in Switzerland for joint programs. As pilot project for the proposed research and training centre, the Institute of Spatial and Landscape Planning (IRL) of the ETH, formerly known as the Institute of Local, Regional and National Planning (ORL), initiated a cooperative on sustainability issues for and with cities in Southern Africa. During a meeting in July 2001 at Johannesburg, researchers from the ETH, Witwatersrand University and the University of Botswana, as well as representatives of the *International Council for Local Environmental Initiatives* (ICLEI) and the Municipality Council of Johannesburg were brought together. A concept for the future research and training collaboration was outlined, however, the financing of the project was not clear at the time.

In parallel, the Department of Urban Studies and Planning (DUSP) at MIT, based on previous experiences, had the idea to carry out an urban design studio on Santiago de Chile in collaboration with the Pontificia Universidad Católica de Chile (Catholic University of Chile) to assess the challenges and potentials for sustainable urban development in a neighborhood of the capital of Chile.

Both groups from the ETH and MIT contributed significant relevant experience in environmental urban planning and the sustainable development of cities. Having already carried out significant research projects on sustainable urban development, both MIT and ETH were in search of competent partners from academia and practice to develop specific case studies. They found these together under the conceptual framework of and contact possibilities enhanced by the AGS.

The opportunity to collaborate under the assignment and with financial support of the AGS proved to be an excellent platform. MIT and ETH staff drafted a first research proposal during a technical meeting in Cambridge, Massachusetts in September 2001, initiating the project *Designing, Implementing, and Measuring Sustainable Urban Development* (DIMSUD). Additional AGS research partners from the Chalmers Environmental Initiative

(CEI) of Chalmers University of Technology joined the project team in the same year. The Institute of Advanced Studies (IAS) of the United Nations University in Tokyo, with broad experiences from its *Mega-City and Urban Development Project,* provided advice and encouragenment while the University of Tokyo had an observatory status throughout the project.

Based on preexisting and newly established research networking connections of the AGS partners to the University of Botswana, the University of Witwatersrand in South Africa, and the Catholic University of Chile, the cities of Gaborone, Johannesburg and Santiago de Chile were chosen as case studies. The selection of these three cities from different regions and with different general characteristics was meant to enable a better understanding of the commonalities and differences relating to sustainable urban development motors, challenges and potential solutions. These cities have quite different historical and socioeconomic backgrounds. Although they are at different stages of development, they all face serious problems with regard to the negative effects of rapid, massive urban development (see Chapter 3). In all, they epitomize medium-sized, rapidly growing primary cities in developing and transition countries on the way to becoming mega cities (Santiago, Johannesburg) or million cities (Gaborone).

Table 2-1. Key geographic, demographic and economic attributes of the case study cities

	Johannesburg	Gaborone	Santiago de Chile
Latitude	26.19°S	24.65°S	33.46°S
Longitude	28.04° E	25.91°E	70.64°W
Population (2004)	3.2 million (City of Johannesburg) 8.8 million (Province of Gauteng)	199,600 (City of Gaborone) 270,000 (Greater Gaborone Area)	1.44 million (Santiago comuna) 4.43 million (Metropolitana) 5.41 million (Greater Santiago)
Average annual population growth rate	0.9% (2000–2010)	3.4% (1991–2001)	1.43% (1992–2002)
Total area	2,300 km^2	970 km^2 (Greater Gaborone) 169km^2 (Gaborone City)	620 km^2 (Greater Santiago)
Municipal budget	R9 billion (US$ 1.422 billion US$)	P180 million (US$ 38.5 million)	

The final proposal for the DIMSUD project (available on http://sustainability.ethz.ch) was submitted to the AGS in October 2001. After approval and generous sponsoring, the project kickoff took place during the AGS annual meeting in Costa Rica in March 2002.

2. THE DIMSUD METHODOLOGY

2.1 An Innovative Approach for Collaboration

An initial challenge before starting the project was the design of a method for collaboration. How can we optimize research collaboration among various partners from different countries, with different cultures and scientific experiences? Velho and Velho (1996) developed an ideal concept for research collaboration with the following characteristics:

- Equally divided tasks
- Common sources of support
- Pooled intellectual effort
- Shared inputs that lead to shared outputs, for example, coauthored reports and publications

Indeed, sustainable development issues require the involvement of all stakeholders. Scientists of various netted and integrated disciplines must be able to take part in this process, as they are the bearers of change and innovation through their research and teaching. Moreover, it is important that they assist public decision makers in choosing the most effective and suitable solutions for each case study city. Establishing a platform for interactions with regard to sustainable urban development fulfilled this task. The collaboration in the DIMSUD project was characterized by a multidisciplinary and participatory approach that combined research, urban design, and capacity building. This platform spanned cities, regions, and universities on five continents (see Figure 2-1).

To implement true collaboration, and not just scientific imperialism of researchers from the North (Dickson 1988), cooperation was based on the principle of equity, which translated into the equal distribution of research funds and shared responsibility for the elaboration of research results and publication. Ultimately, the DIMSUD network of knowledge between several institutes of AGS member universities and non-AGS members successfully promoted sustainable development through South-South and South-North partnerships.

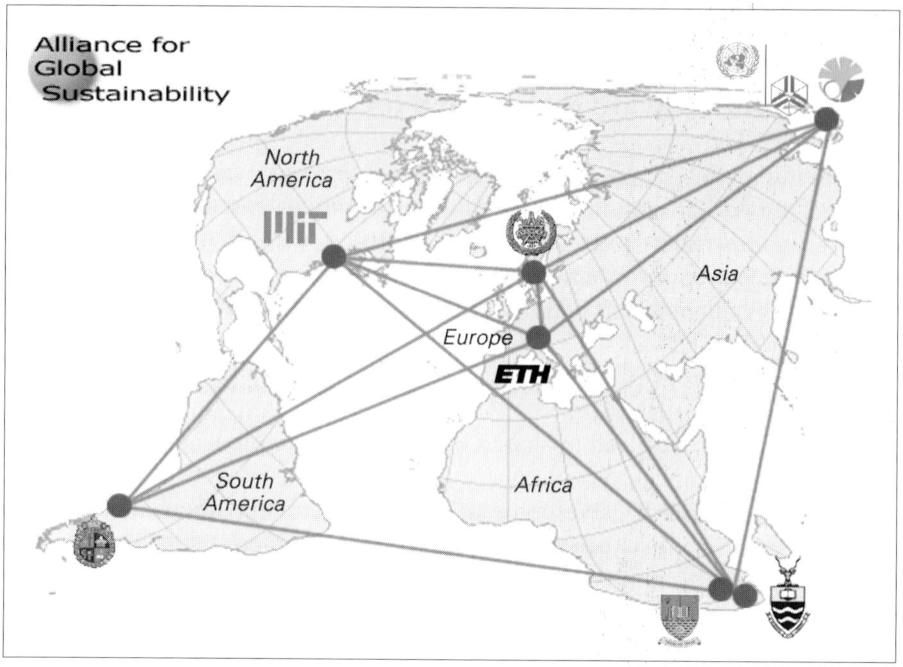

Figure 2-1. The DIMSUD project network

In order to contribute to the AGS objective "to develop a new generation of leaders for sustainable urban development," the participation of and collaboration with students and officials from the three case study cities was an important transdisciplinary precondition for DIMSUD. Collaboration between and also among researchers and students (postdoctoral researchers, graduates and undergraduates) from different universities took place. Field visits at the early stages of research were indispensable for developing a deeper understanding of the realities in the case study cities. Due to additional external funding by the *Swiss Commission for Research Partnerships with Developing Countries* (KFPE), the participation of young Swiss, American, and Swedish students in fieldwork and urban design studios (see Section 2.2) could be supported. This fieldwork resulted in several diploma and semester theses that were a valuable basis for the work of the senior researchers. Students were able to build a long-term personal commitment to the host countries and gain insight in development problems abroad. In the end, the DIMSUD project created the opportunity to establish long-term partnerships between the involved actors and led to personal friendships and ongoing collaboration between the partners.

The project also aimed to bridge the gap between academia and practice. Moreover, an effort was made to reduce the traditional disconnect between the key disciplines of architecture, planning, and economics. Practical research and urban design methods were combined to enable sustainable design solutions that were effectively paired with the mechanisms necessary for their implementation and the tools for analyzing their ultimate impacts (MIT-CDD 2003). DIMSUD is also linked to post-project training on sustainable urban development issues for local decision makers in the case study cities (see Section 2.4) and contains a dissemination component so that a base for continuous learning and application exists. Figure 2-2 illustrates the interactions supported by the DIMSUD project.

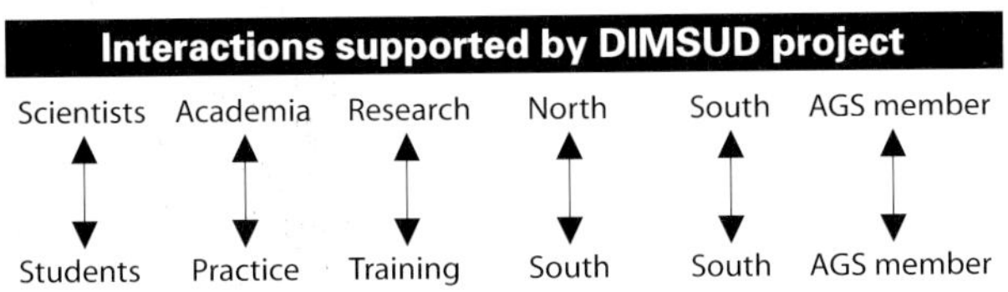

Figure 2-2. Interactions of the DIMSUD project

2.2 Research Themes and Urban Design Studios

The key steps in the initial stages of the project included establishing clear scientific goals, ensuring stable funding from the AGS, addressing asymmetries in resources, promoting equity in partnerships as well as defining the project design, management structure, and approach for information dissemination. In order to achieve DIMSUD's main goal of implementing the abstract concept of sustainable development in specific urban environments, the project was designed at two levels. A macro level methodological framework was used to ensure a general consistency in the project approach and comparability of the results. The project also allowed for some degree of micro level flexibility in its implementation to ensure that the results were geared to match local capabilities and respond to local needs and realities. The project effectively combined collaborative research, fieldwork, and two graduate level urban design studio courses, which involved researchers, faculty, and students from all participating universities.

One of DIMSUD's main tasks was the comparative synthesis of research results across the case study cities grouped under four themes:

- Challenges and opportunities of sustainable urban development
- Strategies, tools, and mechanisms for sustainable urban development
- Analytical tools and models for sustainable urban development
- Indicators in sustainable and integrated planning

In order to efficiently achieve scientific results, bilateral partnerships were established for each case study city (see Figure 2-3).

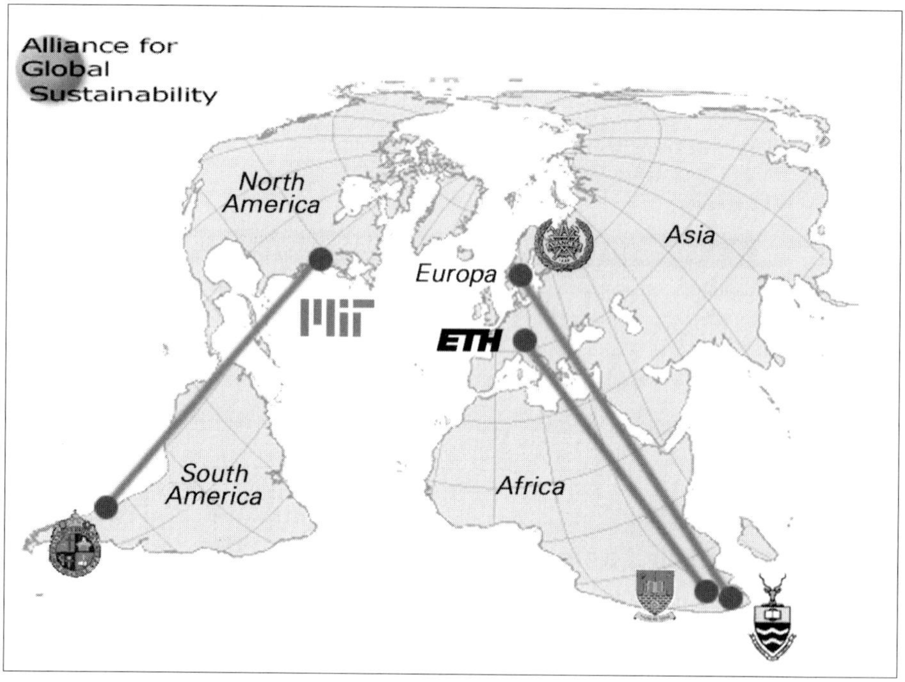

Figure 2-3. DIMSUD: Bilateral research teams

The bilateral work on the four research themes resulted in three case study reports (DIMSUD working papers, downloadable on http://sustainability.ethz.ch). In a next step, each research theme was synthesized across the three case study cities, resulting in individual chapters, which are compiled in this book. Figure 2-4 illustrates this methodology.

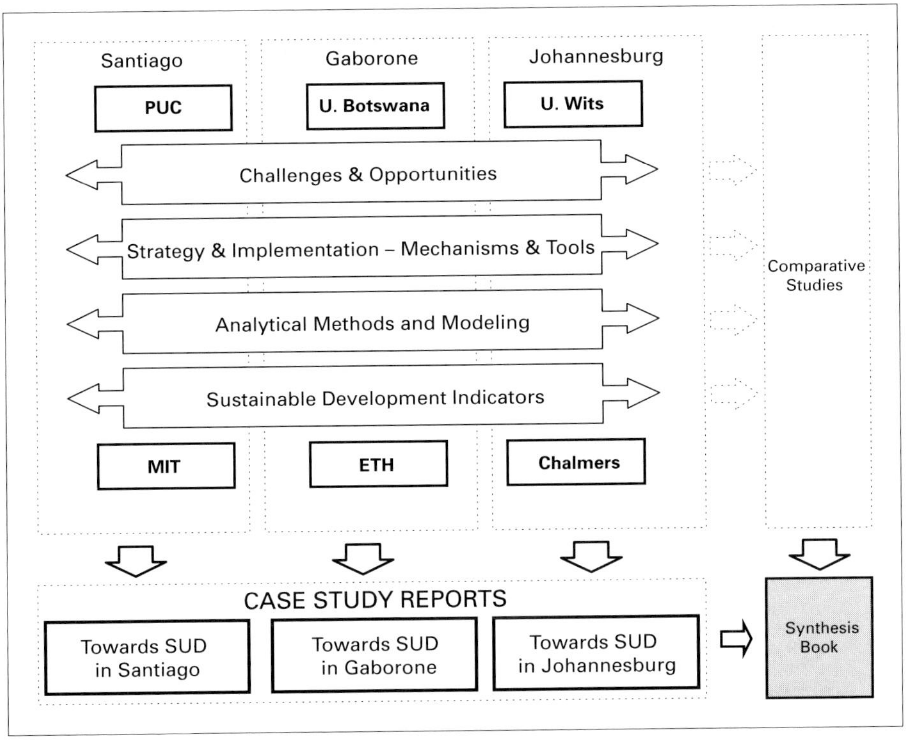

Figure 2-4. Framework for research collaboration and synthesis: City teams, crosscutting themes and outputs

Challenges and opportunities of sustainable urban development

Based in part on the analysis leading to the development of indicators (see below), research in this area focused on internal and external driving forces impacting sustainability in each of the urban case study areas. Research in each city concentrated on the challenges and opportunities of sustainable urban development (see Chapters 4 and 5) along several lines:

- History and geographical context
- Economy
- Environmental attitudes and policies
- Demographics and social issues
- Public finance and investment
- Institutional capacity, government structure, politics, and legal framework

Strategies, tools, and mechanisms for sustainable urban development

This research theme identified existing mechanisms and tools and developed new ones that can be employed to move urban development towards sustainability. Focus was given to instruments and procedures in public finance, participation and planning currently used in the case study cities. These instruments and mechanisms were then refined in order to better guide and maintain sustainable urban development. The different case study cities were able to take advantage of great opportunities to learn from each other in this phase.

Analytical tools and models for sustainable urban development

Models and other analytical tools (e.g. for monitoring and controlling urban development) were developed using key sustainability indicators. These tools were capable of measuring the impacts of the proposed designs and implementation mechanisms. The focus of research was mainly on land use (especially settlement patterns, housing and environment) and transportation modeling and required assessments for each city:

- Capability assessment: inventory of the tools used, transportation models, land use models and integrated models
- Use assessment: What are the tools actually used for? By whom? With what results?

These assessments gave a basis for improving the current modeling capabilities in each city and identified opportunities for cross learning among the cities.

Indicators in sustainable and integrated planning

Although undeniably controversial in attempting to simplify the complex concept of sustainability, indicators can provide a representation of the current state and thus offer a benchmark by which progress might be measured (see Chapter 6). Researchers at each of the responsible institutions have worked to develop relevant sustainable urban development indicators for their respective cities. The indicators were used to monitor the efficiency and effectiveness of strategies, plans, policies, and designs with regard to sustainable urban development.

Research on these four topics was conducted in Gaborone, Santiago, and Johannesburg in collaboration with a local university. Complementary research for each case city was also undertaken at the AGS member universities. Researchers at the local universities gathered relevant information in consultation with city council officials and national authorities through personal interviews and workshops. Additional sources of data, such as official municipal and government documents and websites, policy documents and reports from other stakeholders and academic publications, were identified and reviewed in order to arrive at the preliminary findings of the study:

- Existing data and indicator sets already being used in the case study cities
- Key data gaps
- Institutional and administrative structure of the cities
- Vision and main goals or priorities of the cities

– Key local, national, regional, and global sustainability challenges
– Key tools and mechanisms in city development planning and implementation and their relationship to sustainability
– Analytic and modeling methods currently used in city studies and management processes

Regular consultation with city council officials was used throughout the study for three main purposes:

– To solicit a variety of data and information not explicit in the key documents which capture the cities' vision and development priorities
– To validate and confirm key observations and findings
– To confirm the prime issues and goals of the cities in various sectors in order to ensure relevance of indicators to cities' programs

This work was complemented by a desktop appraisal of existing reports, studies, and policies on the case study cities. The information was compared with findings in scientific literature and international data assessed by the AGS partner universities.

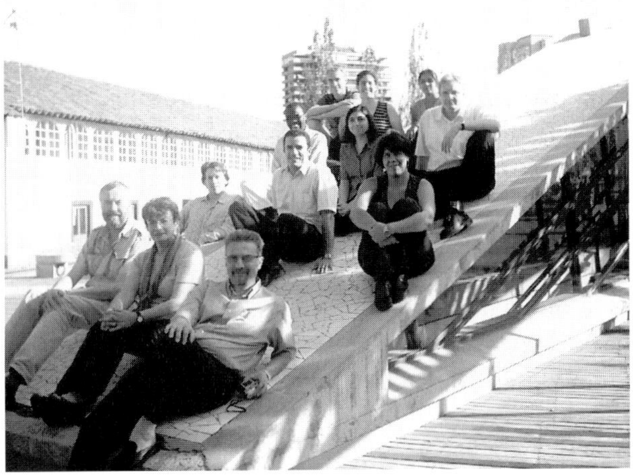

Figure 2-5. DIMSUD researchers during the Santiago urban design studio

In the case of Santiago, an urban design studio was carried out in December 2002 and February 2003, which focused on the neighborhood ("comuna") of San Joaquín through the investigation of the research themes outlined above (see Chapter 6). The studio was built around graduate level courses conducted at MIT. Students came up with specific design proposals during the semester to address urban development problems. Members of the government, the private sector, the nongovernmental community and others participated in the design studio workshops (MIT-CDD 2003).

Based on the experiences and the success of the Santiago urban design studio, Wits and Chalmers Universities decided to implement another studio that took place at Johannesburg in May and June 2003. Students from Chalmers and Wits collaborated on a sustainable urban development project based on the Alexandra Urban Renewal Program. A competition brief for an environmental theme park, a monitoring and evaluation framework for urban agriculture businesses in Alexandra and a top-structure assessment for the regularization of informal buildings in the old Alexandra township were developed as part of the studio. In connection to the urban design studio, a consultation workshop with council officials and other stakeholders in sustainable urban development was held during the May 2003 *Sustainable Buildings Conference* in Pretoria.

2.3 Research Limitations

The DIMSUD project proved to be an effective platform for international collaboration and the synthesis of new ideas for sustainable urban development. However, as in all projects, there are limitations inherent in the research methodology chosen. The first limitation of DIMSUD lies in the hypothesis that the concept of sustainability can be measured and compared. The four research themes discussed above in Section 2.2 are laid out as operational steps in a process that leads to the ultimate goal of sustainability. However, sustainability is not really a "state" that can be reached. Sustainable urban development is a multidimensional concept that can, at least partially, be measured through indicators. "Measuring" sustainability makes the concept easier to understand, which in turn, helps identify the challenges and opportunities for achieving it. As a result, workable strategies for moving towards sustainability may be developed, and using adequate analytical techniques, the impacts of those strategies can be estimated. Therefore, DIMSUD's research focused more on measuring the steps that lead to sustainability instead of measuring the concept or state itself.

Another limitation involves the pre-selection of the case study cities. The six universities involved already had working relationships focusing on the three case study cities. While pre-established relationships provided invaluable contacts and information that improved the quality of the research, the selection of the case study cities might be methodologically debatable (see, e.g. Yin 1994). Other case study cities and partner universities might have been chosen. The question whether a fast growing Asian city should also be used as a case study was raised. Due to limited funding possibilities and the absence of the University of Tokyo as a full project member, the research partners decided to concentrate on the African and Latin American context and eventually extend the project in a second phase to include examples in Asia and the Pacific.

Also, due to time constraints, the DIMSUD project could not entirely achieve its ambitious goals. Particularly the development of analytical tools for steering urban development toward sustainability could have been more extensive. A comprehensive sustainability indicator set could only be worked out for the City of Gaborone, one linked to recommendations of how to implement it through monitoring and controlling (see Bührer 2002, Thalmann 2002 and Keiner and Salmerón 2003). However, based on this example and on the general recommendations given in this book (Chapters 5 to 7), the municipalities of Santiago and Johannesburg will be able to elaborate and implement analogue instruments in ongoing post-project collaboration with their local universities.

2.4 Experiences, Training Needs, and Dissemination

This project intended to leverage a significant number of skills, resources, and tools in order to apply them to a broader challenge: urban sustainability. Important accomplishments included the studio experiences of MIT and Chalmers as well as research on transportation, urban development and the environment, including significant work in Santiago de Chile through MIT's Laboratory for Energy and Environment (LFEE). Other accomplishments included an increased knowledge of the development of urban sustainability indicators for monitoring purposes through research at the ETH, improved methods for public participation and awareness creation for sustainability issues through Chalmers and considerable investment in the development of locally relevant knowledge and skills at each of the local partner universities.

The urban design studios not only furthered their own use as an educational tool but were also integrated into the regular curriculum of the other partner universities. The collaborative nature of the work leading up to and including the design studios offers a good model for future AGS activity.

Training is an important component of international cooperation. Training activities encourage young researchers to become future leaders in science and improve the ability of decision makers and technical staff in planning authorities to identify and successfully implement strategies and measures that lead to more sustainable development. In the post-project phase, the training component of the DIMSUD project will be carried out in collaboration with the United Nations University and the *International Council for Local Environmental Initiatives* (ICLEI), which has a capacity building program aimed at local government leaders (see Chapter 7, Section 2). With representation in South America (Rio de Janeiro) and a regional office in

Africa (Johannesburg), ICLEI's training programs will address the needs identified through DIMSUD's research.

In addition to this book, the DIMSUD homepage (http://sustainability.ethz.ch), several coauthored articles (e.g. Keiner and Cavric 2003, Cavric and Keiner 2003, Keiner and Schmid 2003, Keiner 2004a and 2004b, Keiner, Salmerón and Schmid 2004) and workshop panel discussions, an international conference *Sustainability and Urban Growth in Developing Countries* (see http://www.nsl.ethz.ch/irl/dimsud) was held on the Monte Veritá (Ascona, Switzerland) from October 31 to November 3, 2004. This conference provided an opportunity to present the findings of the DIMSUD project to the public and discuss them with internationally acknowledged experts on sustainable urban development in developing countries. The results of this feedback were published (Keiner, Schmid and Koll-Schretzenmayr 2004) and thus made accessible to all interested researchers and practitioners dealing with sustainable urban development.

REFERENCES

AGS 2003. *http://www.globalsustainability.org/*, retrieved on November 20, 2003.

Bührer, C. 2002. *Indicators in Sustainable and Integrated Planning in Gaborone*. Diploma Thesis. Institute for Territorial Development and Landscape, Department of Civil, Environmental and Geomatics Engineering, Swiss Federal Institute of Technology, Zurich.

Cavric, B. and M. Keiner 2003. Toward Sustainable Rural Land Use Planning Practices in Botswana: The Concept for an "Integrated Rural Plan." *ETH e-Collection*. http://e-collection.ethbib.ethz.ch/cgi-bin/show.pl?type=bericht&nr=322&lang=en, retrieved on May 24, 2004.

Dickson, D. 1988. Knowledge as Imperialism. In D. Dickson (Ed.) *Science and Foreign Policy: The New Politics of Science*. Chicago: The University of Chicago Press.

Greene, M. and M. Johnston 1997. *Contradictions in Collaboration: New Thinking on School/University Partnerships*. New York: Teachers College Press.

Jonas, H. 1985. *The Imperative of Responsibility: In Search of an Ethic for the Technological Age*. University of Chicago Press.

Keiner, M. 2000. Auf dem Weg zu einer Erdcharta. *collage*, Vol. 4: 32–34.

Keiner, M. 2004a. Sustainable Development and Urban Management in Developing Countries: The Case of Africa. In M. Koll-Schretzenmayr, M. Keiner, and G. Nussbaumer (Eds.): *The Real and the Virtual Worlds of Spatial Planning*. Berlin, Heidelberg: Springer: 43–59.

Keiner, M. 2004b. Toward Gigapolis? From Urban Growth to Evolutionable Sub-Mega Cities. In M. Keiner, W.A. Schmid, and M. Koll-Schretzenmayr (Eds.) *Managing Urban Futures: Sustainability and Urban Growth in Developing Countries*. Aldershot: Ashgate Publishers (forthcoming).

Keiner, M. and B. Cavric 2003. Managing the Development of a Fast Growing City: Gaborone, Botswana. *ETH e*-Collection. http://e-collection.ethbib.ethz.ch/cgi-bin/show.pl?type= bericht&nr=321&lang=en, retrieved on May 24, 2004.

Keiner, M. and D. Salmerón 2003. Sustainable Urban Development in Gaborone, Botswana. DIMSUD Working Paper. *ETH e-Collection*. http://e-collection.ethbib.ethz.ch/cgi-bin/show.pl?type=bericht&nr=308, retrieved on May 6, 2004.

Keiner, M., D. Salmerón and W.A. Schmid 2004. Nachhaltige Stadtentwicklung in Entwicklungsländern. *Studio!Sus*, SS04, No. 2: 13–15.

Keiner, M. and W.A. Schmid 2003. Urbanisierungstendenzen in Entwicklungsländern – Probleme und Potenziale für nachhaltige Stadtentwicklung. *DISP* 155: 49–56.

Keiner, M., W.A. Schmid, and M. Koll-Schretzenmayr (Eds.) 2004. *Managing Urban Futures: Sustainability and Urban Growth in Developing Countries*. Aldershot: Ashgate Publishers (forthcoming).

Labarca, A. and J.M. Hendrickson 1999. *Our Global Village*. Boston, MA: Heinle & Heinle.

MIT-CDD 2003. *Santiago Urban Planning Studio: Sustainable San Joaquín*. City Design and Development Group, Department of Urban Studies and Planning, Massachusetts Institute of Technology, Cambridge, MA.

Moavenzadeh, F., K. Hanaki, and P. Baccini (Eds.) 2002. *Future Cities: Dynamics and Sustainability*. Alliance for Global Sustainability book series, Vol. 1. Dordrecht, Boston, London: Kluwer Academic Publishers.

Shuman, M. 1994. *Towards a Global Village: International Community Development Initiatives*. London: Pluto Press.

Staub, S. and P. Green (Eds.) 1992. *Psychology and Social Responsibility: Facing Global Challenges*. New York University Press.

Thalmann, B. 2002. *Analytical Tools and Models for Sustainable Urban Development – A Framework for Spatial Monitoring and Controlling in Gaborone (Botswana)*. Diploma Thesis, diploma course Rural Engineering, ETH Zurich.

UNESCO 1998. *World Science Report 1998*. Paris.

Velho, L. and P. Velho 1996. Scientific Collaboration of Advanced/Developing Countries in Biological Sciences: The Case of the Maraca Rain Forest Project. *Cadernos de Ciência & Tecnologia*, Brasília, Vol. 13, No. 1: 9–20.

White, F. 1998. *The Overview Effect: Space Exploration and Human Evolution*. Second Edition. Reston, VA: American Institute of Aeronautics and Astronautics.

Yin, R.K. 1994. *Case Study Research: Design and Methods*. Third Edition. Thousand Oaks, CA: Sage Publications.

Chapter 3

THE CASE STUDY CITIES

Santiago de Chile, Johannesburg, Gaborone

Branko I. Cavric[1], Aloyse C. Mosha[1], Marco Keiner[2]
[1]University of Botswana, [2]ETH Zurich

Abstract: This chapter examines the "sense of place" of Santiago de Chile, Johannesburg and Gaborone, reflecting on their environmental, physical, socioeconomic, policy and development perspectives. In each individual case study city, short textual episodes and topical themes are used to focus on the distinctive urban settings and sustainability experiences of previous and newly born generations of Latin American and African urban dwellers. For this chapter, the background information is resourced from individual DIMSUD project city reports covering a wide range of urban narrations and descriptions aiming at supporting more sustainable urban practices. After two years of investigation by DIMSUD partners, this chapter draws on synthetic rendering to try to come closer to understanding the individual qualities and dynamics of these three rapidly expanding cities.

Key words: Latin American city, urban morphology, sustainability, apartheid, urban environment, integrated development planning

1. SANTIAGO DE CHILE – A CITY OF RHYTHMS

1.1 Urban Development after the Spanish Conquest

In order to understand the factors influencing the urban development of Santiago de Chile, it is necessary to understand the influence of the following historical periods: the "mercantilist" colonial period (1500–1750), the period of "outward growth" (1750–1914), the crisis of the "liberal model" period (1914–1950) and the current "period of transnational capitalism" (Galafassi 2002, Violich and Daughters 1987).

M. Keiner et al. (eds.), From Understanding to Action, 41–73.

The origins of contemporary Santiago City planning are to be found in the Spanish Law of Indies (1573). Initially, the small village borne next to the Mapocho River and at the foot of the Huelén Hill began to extend towards the surrounding areas, first in an "organic" and later on in a "planned" manner.

Towards the end of the century of the Conquest and upon the arrival of Governor García Hurtado de Mendoza, Santiago became the heart of the administration, competing with the city of Concepción, the military center (Barros, Inostroza and Allison 2002). The characteristics of this period up to the beginning of the Industrial Era point to close ties between tradition and progressive developments. Spontaneous urban segments became interlocked into the more regularly defined new quarters. Through the process of continuous metamorphosis, a new geometry was imposed over informally planned semicircular enclaves of the original settlement, gradually transforming the village into a more permanent urban milieu. Colonial Spanish planning left the old areas with important buildings and other structures intact, spreading an unvarying grid over 27 km^2 of the outlying terrain of the Mapocho Valley and its intruding hills (Barros, Inostroza and Allison 2002).

During the time of the "mercantilist" colonial period, when the expansion and changes within the historic core started to change the city's image, the concept of a restored gridiron, based on balancing the new developments with restoration and conservation activities, tailored the city's urban mosaic. The central parts of the city, designated for governmental institutions and elite residences, were punctuated with parks, plazas, squares, avenues, and religious, educational and other ancillary uses. The reorganization of blocks allowed the development of different building types, changing densities in a more or less very stable manner. Fueled by a mining boom during the early years of the Republic – founded in 1810 – and numerous ideas of fine thinkers, scholars and administrators, Santiago continued to produce fragments of elegant and sustainable urban design with clear expressions of sensitive perspectives and well-balanced proportions.

During the late 18th and at the beginning of 19th century, Santiago grew rapidly and acquired the first ring of poorer districts surrounding the Belt Road created by Vicuña Mackenna. It was the beginning of a new era when emerging industrial activities transformed the urban landscape and changed the life of people living in traditional residential quarters. At the same time, this was the period of "outward growth" when the city was dependent on primary exports and the processing of local and imported raw materials.

Industrialization created a new urban form essentially by dividing the urban fabric into special districts based on differentiated land values and transformations of ownership. An influx of new immigrants wishing to come and live in the city was inevitable, and the city government reacted by encouraging the segregation of various land uses following different land-value gradients. The streets, which continue the original grid lines from the 17th century, were extended further in very long stretches, adding new opportunities for city expansion.

The plan of Santiago de Chile of that time had a cross-axial structure with numerous interconnected rectangular and round plazas, surrounded by blocks developed on a grid. This planning pattern was about bringing Santiago's numerous districts a greater sense of connectivity and coherence towards neighboring settlements and regions. However, the grid was not always a good framework for socially equitable development and increased socio-political tensions among rich and deprived city dwellers. This crisis of the "liberal city model" and the current period of "transnational capitalism" and "architectural modernism" influenced the development of newer strategies in reshaping the city.

1.2 An Environment of Distinctiveness

The City of Santiago de Chile is the central hub of a rapidly expanding metropolitan region better known as Metropolitana de Santiago or "Greater Santiago." Greater Santiago consists of 62 km^2 of continuous urban development. Greater Santiago is the biggest and most important city in the country with more than 5 million people – which is one-third of Chile's population – living in its 34 *comunas*.

Much of Santiago's urban history, economy, landscapes and development have been closely related to the proximity of the Andes and Pacific mountain façade.

Through initial tectonic and volcanic activity and later glacial, eolian and fluvial erosion, the landscape has amalgamated into three dominant physiographic units, namely the Andean massif, hills of La Dehesa and Santa María de Manquehue, and the meandering Mapocho River with its tributaries Maipo, Puangue, Lampa, Colina and Angostura.

The most common characteristic of the city's climate is how it has four distinctive seasons with sunny springs, hot summers, cool autumns, and rainy winters. Virtually all the rain falls during May, June, July and August, the winter months in the southern hemisphere. The average period of daily sunshine is more than ten hours, even in winter, and temperatures are highest and most uniform during the summer months.

The unique urban fabric of Santiago has been developed through continuous human habitation and organization of different activities since the city's beginnings. These two elements of urban change and land use dynamics have created an elaborated matrix of extensive artificial patterns resulting initially from deforestation and various agricultural practices, and later from intensified processes of industrialization and urbanization. Centuries of Santiago's expansion have sprawled boundaries of its indigenous environmental setting, affecting radical changes in the city's natural biodiversity and environmental systems. In this evolutionary journey, maintaining urban land resources of Santiago's agglomeration have become a priority and subject to the competitive land-use game among numerous stakeholders with their planned activities, needs, wishes, speculations and priorities. With regard to the city's environmental framework, future patterns of Santiago's urban development must be examined in relation to the degree of pollution and exploitation of the available natural resource base (see Chapter 4, Section 8).

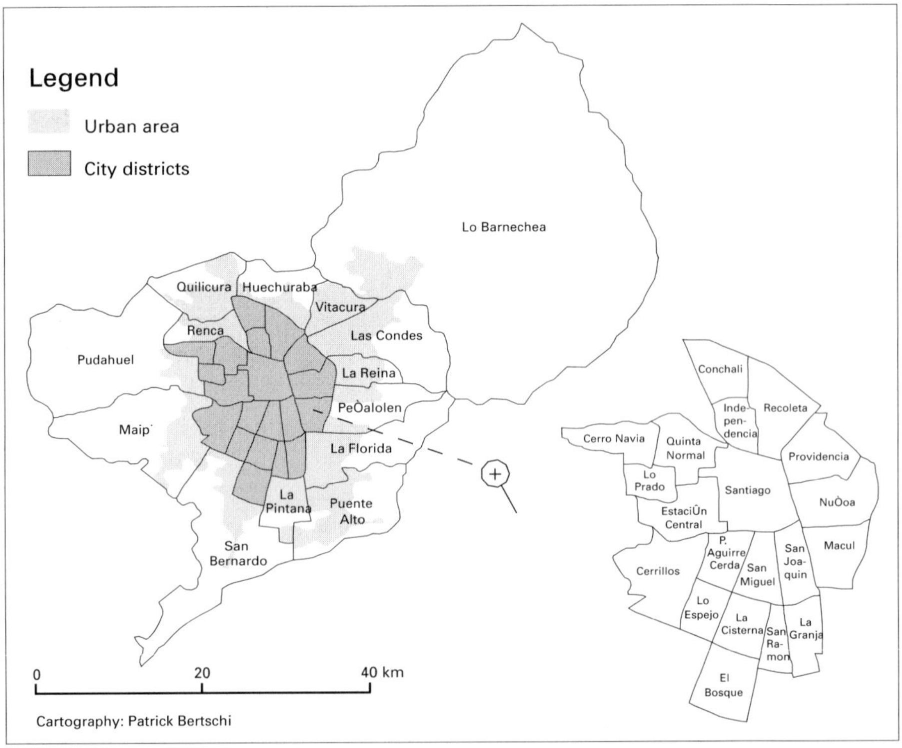

Figure 3-1. Administrative boundaries of Santiago de Chile

1.3 Santiago Contemporary Morphology

Today, the city's planners are making efforts to curb and soften the impacts of Santiago's historical grid by imposing gently undulating major transportation corridors that are accented with interesting landscaping arrangements and incidental irregularities to bring more variation into the urban landscape (e.g. Mapocho River corridor). In addition, mixing land uses and their socioeconomic impacts lead to the creation of more sympathetic cityscapes in which Santiago maintains its image as a traditional and modernist city, where development projects requiring massive government and private investors' intervention and coordination are shaping the contemporary city morphology.

The general picture of Santiago's historical and environmental urban trajectory radically changed over the years, highlighting the city's distinctive multicultural mix and major structural transformations that occurred particularly in the last 35 years. The urban structure of Santiago comes close to the general *New and Improved Latin American City Structure Land Use Model* (Ford 1996, see also Griffin and Ford 1980 and 1983, Ford 1996, Bromley and Jones 1996, Crowley 1995). This model consists of the following elements:

- A downtown area as central business district (market, plaza)
- A commercial spine along the main streets leading up to the CBD housing the city's most important economic, social, and cultural amenities, and associated elite residential sector since the colonial times as well as a zone of maturity, sometimes characterized by gentrification
- A series of concentric and semi concentric zones (zones of in situ accretion, zones of peripheral squatter settlements, called "periférico" or "poblaciones callampas;" mushroom towns) in which residential quality, infrastructure and public services decrease with distance from the city centers.

Global convergence in consumer and residential patterns influenced the division of Santiago's central area (Santiago, Providencia, Recoleta, and Independencia districts) into a high-tech and vibrant Central Business District (CBD) with Manhattan Criollo and El Golf; and the Mall with administrative and commercial core surrounding La Moneda. There are also several additional enclaves of traditional markets. All of these areas are linked by pedestrian and public transport corridors developed in the form of tree-lined boulevards. In the immediate vicinity of the CBD urban ring, rich residential neighborhoods dominate, like Los Leones, Providencia, Bellavista, and others.

Santiago's CBD skyline looks similar to that of other Latin American and global cities due to modern high-rise office blocks, impressive multinational headquarters, and the prominence of western retailers (Potter and Lloyd-Evans 1998). Increasingly, upper middle class apartments and houses are being built on the fringes of the elite residential sector, in part to satisfy increasing demand in this market and to act as a buffer against the poorer residential settlements nearby (Pacione 2001).

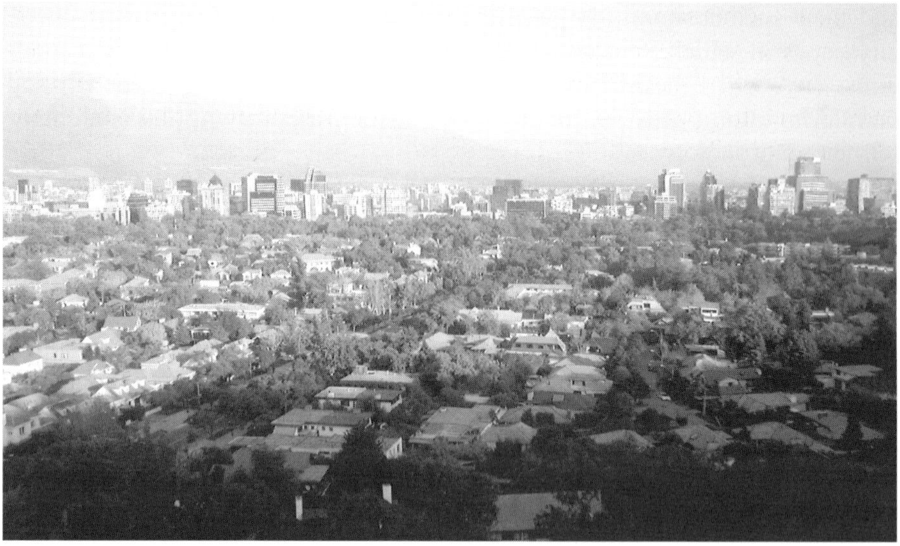

Figure 3-2. Santiago's skyline and residential areas (Photo: B. Cavric)

A series of different zones shape the outer edge of the CBD and central commercial spine where the quality of life, socioeconomic opportunities, and the level of facility and utility services decline with the distance from the inner city areas. The pattern of urban space here is tailored through the integration of industrial, transitional and specific residential land use zones (e.g. zone of maturity, zone in situ accretion, middle-class residential zones and zone of squatter settlements or *callampas* such as Maipú, San Bernardo, and La Pintana). Industrial zones are developed following railway and highway lines (Quilicura, Quinta Normal Cerbilos, San Bernardo, San Miguel, San Joaquin, Puente Alto, San Ramon, etc).

As with any metropolitan area, Santiago is experiencing a process of urban growth that obeys the dynamics of supply and demand for land and mobility, which in turn, are regulated by the State through urban instruments and policies. Primacy, prominence and rapid growth based on rural-urban migration are basic characteristics of Santiago as the capital city of Chile.

From the point of view of demand, it can be highlighted that the contemporary urban pattern of Santiago has resulted from the migrations of people from "less-urban" to "more-urban areas" as well as to a natural increase in the already urbanized population (Redclift 1973) during the industrialization period (1930–1950). In a second period (1980–2000) modifications to the pattern of income occurred which notably increased the demand for housing and infrastructure, especially in peripheral areas. At the same time, the opening of the country to the exterior and the liberalization of markets influenced changes in the metropolitan economic base, increasing demand for services and new industrial poles.

From the point of view of supply, the growth of Santiago can be explained as a sequential process, involving a search for urban income and appreciation led by market players and public housing programs and the city's subsequent occupation by people according to their payment and debt capacity. This trend generated territorial and socioeconomic segmentation, which impacted land prices, which in turn permitted or prevented the establishment of new households. This also explains the pressure to develop projects outside the city limits in large-scale formats (mega-projects and condominiums), which also applies to public housing plans. Demand and income defined the size of the market, and consequently, the quantification of supplies and their socioeconomic specialization for non-residential uses like retail, services, and private education.

In the period from 1940–1970, expansion was linear along a main road connecting the historical center of Santiago with the commune of Providencia. Trade and commercial activities were also developed on longitudinal roads that connected the centers with outlying residential centers of dynamic growth like Vitacura, San Miguel or Ñuñoa. Only sectors where the density of consumption was relevant experienced an increase in the location of service activities, with projects for offices, hotels, etc.

The consolidation of these sectors triggered demand for residential areas. Due to the high prices of land, large apartment blocks with high densities were developed. This explains why most apartments are concentrated in old suburban neighborhoods that were once absorbed by commercial and service centers. This concentration is directly proportional to the original income of the old neighborhoods, which explains the development of dynamic districts in Providencia, El Golf or San Miguel.

In summary, the following four phenomena can explain the general features of urban growth and urban patterns in contemporary Santiago:

- First, expansive processes are generated by the demand for land and infrastructure, where the city increases in size or density. In this case, the real estate market and the State determine the manner in which peripheral and central territories are developed and later urbanized. This means that in the end, Santiago grows with the sum of disconnected pieces of real estate.
- Second, average income determines the socioeconomic configuration of housing units. This affects patterns of segregation and the attraction of new projects. This variable also explains the densification of certain sectors with large and medium sized buildings (formation of building corridors) and the transformation and tri-partite evolution of land use, with the appearance of financial and service centers that reach from the central area to each expansion pole.

- The third phenomenon is how public policies on territorial zoning and infrastructure interact with market variables. Their impact on the configuration of urban growth has been less in the recent history of Santiago since they generally adapt to the requirements of the market. The exception are the roads connecting the new suburban areas with the main centers, which have the potential to open up and commercialize new rural territories, making them viable for urban expansion.
- The fourth phenomenon is related to the socioeconomic conversion of outlying and central territories, which started with the first subdivisions of rural residential properties in the southwest sector in the early 1980's. The most exemplary case of this kind of real estate development are the communities of Peñalolen and Huechuraba. The same logic applies to new suburban residential mega projects, which radically change the pattern of income of rural localities, operating as new medium and high-income neighborhoods.

1.4 People of Multicultural Society

The urban dwellers of Santiago possess very distinctive ethnic characteristics. The Spanish and indigenous origins are dominant in the demographic mix, but successive influences of immigrants, mainly from Germany, Arab countries, Israel, Croatia, Italy, Switzerland, and some others have also helped to enrich the Santiago ethnical melting pot. However at the present time, the inhabitants of Santiago, irrespective of their ethnic origins, have a distinctly urban outlook, which is quite unique in Latin America (Verhasselt Puppinck 1997).

At the beginning of the 20th Century, Santiago's size was due to the city's role as Chile's administrative center. By the mid-1940s, the macro-economic policy of import substitution further contributed to Santiago's population dominance, as industrial investments, followed by financial and commercial activities, focused on the city (Errázuriz 1987).

During the first half of the 20th Century, over half of the Santiago Metropolitan Region's (SMR) population growth was due to migration, a share which gradually declined to 20% during the 1970's to an estimated 8 to 10% today. Census data indicates that population growth – mirroring national patterns – has slowed in recent years, declining from 2.62% per year from 1970–1982 to 1.97% per year from 1980–1992 (national rates were 2.03% and 1.64%, respectively) reaching an annual rate of 1.43% from 1992–2002 (INE 2002). Figure 3-3 illustrates the spatial aspect of population density.

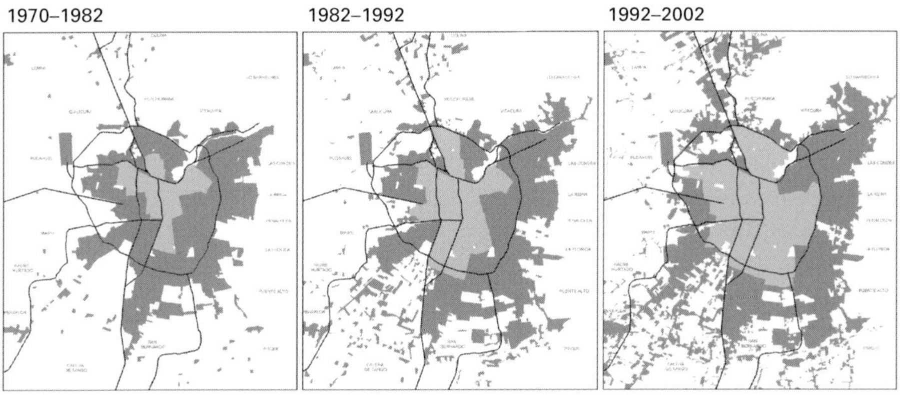

Figure 3-3. The development of urban sprawl in Santiago over three decades. Light grey areas indicate zones with decreasing population density. (Source: Catholic University of Chile 2004)

While the center of the city and the first ring of communes remain the most densely populated areas, the population loss in this part of the SMR has also translated into lowered densities (see Figure 3-4). The second and third rings, in contrast, have shown an increase in population densities in the period 1970–1998, although these *comunas* still display lower population densities on average than the more central areas of the city, producing an archetypical density gradient that declines with distance from the city center.

A convergence of factors has contributed to these population dynamics: low land prices on the urban periphery, which have focused government public housing efforts there, the desire of middle and upper income groups to escape the deteriorating center of the city and live on larger lots in larger homes that are also more financially feasible on the urban periphery; and importantly, a fundamental change in land use in some of the central *comunas* (particularly Santiago and Providencia), where commercial and service uses are replacing formerly residential uses. In general, the increased density of the periphery was basically due to the development of social housing, which in the period 1992–2002 represented the main form of occupancy in Puente Alto and San Bernardo.

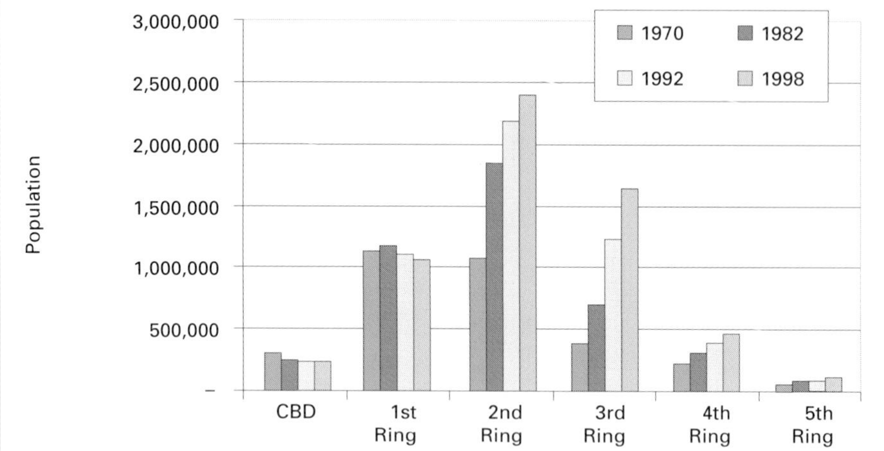

Note: Each ring is comprised of comunas within the following approximate radial distance from CBD (in km from rough geographic center of comuna) – 1st Ring, <5; 2nd Ring, 5–10; 3rd Ring, 10–15; 4th Ring, 25–45 ; 5th Ring, 50–100.

Figure 3-4. Population growth in the Santiago Metropolitan Region by Ring (Source: Zegras and Gakenheimer 2000)

According to census data (INE 2002), the average household size in the Metropolitan Region in 1982 was 4.82 persons, with a range across the *comunas* of 3.5 to 5.7. In general, the lower income communes such as Lo Espejo and La Pintana had the largest household size, while the smallest household sizes were recorded in Providencia and Santiago communes – attributable to the fact that these communes have relatively high incomes and a large concentration of older residents. By 1992, the average household size in the Metropolitan Region of Santiago had declined to 4.25, with a range across communes of 3 to 4.8 persons per home. By 2002, this indicator dropped further to 3.7 persons per home. All communes experienced declining household sizes in part due to depopulation and fragmentation of the home and to new residential demand. For example, the market of single people or young couples without children explains the low indicator of number of persons per home.

1.5 Economy of new Demands

The major weakness of today's Chilean economy is its dependence on the United States as its single largest trading partner. In addition, there are some other constraints like the vulnerability of copper revenues, representing 40% of country's exports and low world market prices. Chile's economy has been subject to competing ideologies for a long time. Allende's socialism brought huge corporations into the state sector. The Pinochet dictatorship introduced radical monetarist policies. Drastic cutting of the state sector and the selling-off of state enterprises at below market value led to large profits for investors and speculators. Tough economic measures, irrespective of the social consequences, brought Chile's inflation rate down from 400% (Heritage 2003). According to the *2004 Index of Economic Freedom* (Heritage Foundation 2004), Chile has the freest economy in Latin America and the Caribbean.

Today, the Santiago Metropolitan Region (SMR) plays a fundamental and continuously growing role in the Chilean economy despite government intentions over the years to decentralize economic growth. Between 1986 and 1996, the economy of the SMR grew at an average annual rate of 7.9% per year, compared to the nation's 6.8% growth rate. The Santiago Region's contribution to the Gross Domestic Product (GDP) increased from 42% to 47% during that period (Central Bank 1999).

Several facts help to highlight the importance of the SMR to the national economy. All foreign and national banks operating in the country and 46 of the nation's 47 major economic groups are headquartered in the SMR. The SMR accounts for 58% of national industrial employment. In 1993, the SMR accounted for 53% of the nation's housing stock and 55% of industrial and service building stock (De Mattos 1996). In addition, the SMR contains a majority of the country's most important centers of higher education. In terms of sectoral contribution to GDP, from 1986 to 1996, the SMR's role in the nation's industrial activity increased from 44% to 52%. Its role in national construction activity increased from 37% to 41%, and its commercial activity increased from 65% to 68%. The leading economic sectors – based on contribution to Gross Regional Product (GRP) – were commercial services, financial services, and industry, accounting for 30%, 27%, and 21% of GRP, respectively (Central Bank 1999).

Although, transportation/telecommunications and construction sectors comprise a relatively small share of economic activity (in 1996, 10% and 5.5% of GRP, respectively), they have proven to be the most dynamic sectors in the economy of the SMR, growing at an average annual rate of 9.8% and 9.6% respectively from 1986–1996. Commercial services followed close behind at an average rate of 9.5%. In terms of employment, government and social services are the leading sectors, accounting for 28% of all jobs, followed by industry (23%) and commercial activities (20%). Together, these sectors provided 1.5 million jobs in 1992, out of a total SMR workforce of slightly more than 2 million (GoRe 1995). Despite the presence of large industrial establishments, small and mid-sized enterprises still dominate industrial employment.

The 1997 economic crisis in Asia had strong repercussions on the Chilean economy as Chile's main export countries, besides the USA, are Japan and China. From 1997 to 1998, economic growth slowed down to 3.4%. Between 1998 and 1999 the slowdown had turned into contraction, and the country experienced a recession (–1.1%); the most adversely affected sector was construction, which shrunk to 10% (Central Bank 2000). The SMR was particularly hard hit, as unemployment, which ranged from 6.7% to 7.5% from 1995 to 1998, jumped to 10.5% in 1999 and increased to more than 14% in 2000, which was higher than the national average (Central Bank 2000). From 1999 to 2002, Chile's unemployment rate slowly declined from 10% to around 9.5% (Central Bank 2003).

Despite the economic recession and despite the fact that Santiago did not and still does not have the necessary resources to maintain and improve basic infrastructure and services, the phenomenon of urban growth took place. As unemployment is relatively high, as poverty is growing, as private investments have dropped, and as the norms of neighborliness and coexistence are giving way to various forms of violations, low-income groups face increasing problems of access to housing and urban services. To promote equitable, democratic, efficient, and environmentally sustainable forms of urban development against such a backdrop is not an easy matter (Carrión 2001).

2. JOHANNESBURG – A RAINBOW METROPOLIS

2.1 Before and Beyond the Apartheid Era

The city of Johannesburg in South Africa is situated about 1800 meters above sea level on the plateau often referred to as the *Highveld*. This may well have been the home for the earliest ancestors, and for at least 100,000 years, humans have inhabited the place. The city was established in 1886 following the discovery of the gold reef of the Witwatersrand. Over time, the number of settlers increased and soon the city grew out of the former mining camp (Beall 2002).

Through a period of just over a century, the area has been transformed from hunting and subsistence livelihoods, to commercial farms, a gold-prospecting camp, mining town and a centre of manufacturing to its current status as a regional and global financial-services and trade hub of the continent and the world.

From its founding to the end of the apartheid in South Africa in 1994, the city was racially divided with an inequitable distribution of wealth and social services among the races with demarcated areas for whites, Indians, Coloreds and Africans. Before the shifting of power to ANC, Africans were considered second-class citizens and were not allowed in white areas nor were they allowed to be in the city other than for work, in which case they required passes. After nightfall, they had to retire to their black home areas.

The municipality constitutes an area of 2,300 km^2, of which buildings and infrastructure occupy half. There are still some green reserves with bushland or trees, some wetland but very few and small water bodies. Through intensive care in private and public open spaces, the city suburbs are characterized by the greenery of trees, shrubs and flowers, which has translated to an artificial forest within the surrounding grasslands. Statistics show that while the quality of life of Black residents – measured by the *Human Development Index* (HDI) – is far below that of their white counterparts, the latter are at par with residents of high income developed countries. According to the HDI, it is theoretically better to live in Johannesburg than elsewhere in South Africa or Sub-Saharan Africa (City of Johannesburg 2002b).

Johannesburg today still exhibits stark differences in the provision and distribution of social amenities. These social amenities include hospitals, education facilities, nature reserves, parks, open spaces, recreational areas, security provisions, preservation of places of cultural and historical heritage, cemeteries, etc. Radical inequalities exist between different regions, e.g. the high-income Sandton area and the lower income Alexandra area for instance. These differences have created several challenges to sustainable urban development in Johannesburg, including high levels of crime, fragmented service delivery, limited socioeconomic opportunities and high poverty levels in former black townships and informal settlements. One of the key indicators for improved social services is the percentage reduction of crime.

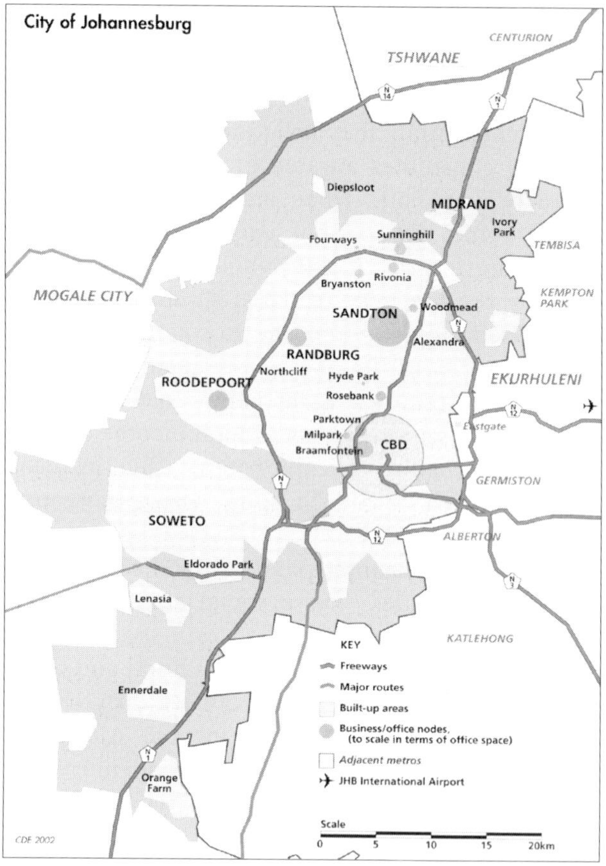

Figure 3-5. City of Johannesburg: Extent and key nodes (Source: Centre for Development and Enterprise 2002)

Since the apartheid era, high qualitative health care has been very unevenly distributed in the city. There is an agglomeration of clinics in the CBD, while the "Deep South" hardly has any (City of Johannesburg 2000b). The Johannesburg General Hospital, with 2000 beds in the city centre, is very modern and well equipped. Baragwanath, situated in Soweto, is said to be the largest hospital of Africa serving 5,000 patients a day, but with very limited resources. Recently, the province launched a new system for primary health care on a district level. The system was designed to address the problem of overcrowding at the public hospitals. Now, there are many local clinics, community health centers and mobile clinics run by the regional administration. There is also a central health information system to provide the planning and logistics. Finally, there are a large number of private hospitals for those who can afford more expensive health-care services.

Education was also provided in a segregated manner with the white population having better education than black residents. Even today, qualitative differences among both private schools and universities are huge (IMC 2003). However, there are ongoing initiatives to standardize the quality of education through the *National Qualification Framework* under the *South African Qualifications Authority*. Other initiatives are also being undertaken to restructure the education system with the introduction of a new curriculum. Low skill levels among Johannesburg's population are highlighted as a severe obstacle to increasing economic growth. To date, 25% of the city's adult population is technically illiterate, another 39% have less than a grade 12 education, and only 4% have a college degree (Allan, Gotz and Joseph 2001). The level of skills in the city is much lower than what is required for the city to operate effectively. The city, and the country as a whole, although attracting skilled labor from other African countries, has experienced a major brain drain in the last decade, as some professionals have reacted to unfavorable and unstable conditions and have emigrated.

Land use management - While apartheid policies encouraged the eradication of illegal settlements, post-apartheid South Africa has been characterized by organized land invasions. On the one hand, legislation such as the *Restitution of Land Act* (RLA); *Extension of Security of Tenure Act* (ESTA) and *Prevention of Illegal Occupation of and Eviction from Land Act* (PIE-Act) creates challenges for developers. On the other hand, there is the issue of pressure on bulk infrastructure, especially from unplanned settlements. Within Johannesburg, various opportunities exist in relation to land use (City of Johannesburg 2002a). They include the upgrading and formalization of informal settlements, development of economic opportunities in close proximity to residential areas, development of local production, support for small, micro and medium enterprises (SMMEs) and labor intensive manufacturing and creating opportunities for tourism and the improvement of public transportation to enhance economic opportunities for the citizens. These opportunities are constrained by urban sprawl, land invasion, limited markets and poor relations between business and residential locations. The key issue that the city needs to track is the impact of these land use patterns on the economy and the environment. This can be tracked by such indicators as the number of employees per sector and average travel time.

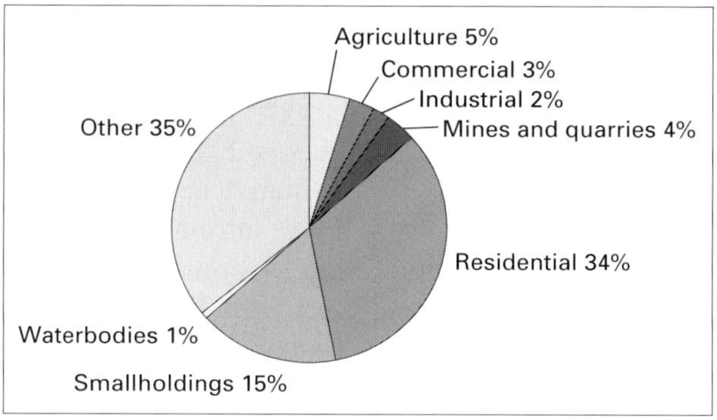

Figure 3-6. Land use types of Johannesburg, percent of total area (Source: Republic of South Africa 2003)

Human settlements - The government of South Africa committed itself to deliver one million low-cost houses in five years through the *Reconstruction and Development Program* (RSA 1994). So far, 1.4 million houses have been delivered (Mahanyele 2002). The predominant mode of delivery has been the one-shot, supply-side, project linked, individualized capital subsidy, whose effectiveness and sustainability has been questioned (Huchzermeyer 2001). The paradigm shift locally is to support the *People's Housing Process*. Whether or not this new paradigm will meet the challenges of sustainable development is still to be seen. Several *Local Integrated Development Plans* (LIDPs; City of Johannesburg 2002c) outline some of the issues of human settlements which challenge sustainable urban development: land invasions, uncontrolled informal settlement growth, lack of employment opportunities in close proximity to existing and proposed housing and lack of an urban boundary.

2.2 People of the African Melting Pot

The Witwatersrand gold reef is characterized by a belt of mines and other industries stretching from east to west through the city of Johannesburg. The old core of the city, the Central Business District, is situated close to where the first mining camp was located on the northern side of the ridge. From this point, the colonial planners laid out the city on a classical grid pattern.

Throughout its history, this great mining and industry city has always attracted migrants from a large hinterland, including other African countries. There are considerable numbers of illegal immigrants, which makes it difficult to determine actual numbers of immigrants and the population in general. According to the 2001 Population Census (Statistics South Africa 2001), the population of the city today stands at 3.2 million people. This is about 7% of the national population (at 44.8 million). The number of households is about 1 million, which means an average of 3.2 persons per household. Over the last 40 years, demographic trends have been towards fewer young children (0–14) and aged (55 upwards) persons with the majority of the population being between 19–39 years old. Population growth is estimated to be slightly over 0.9% (City of Johannesburg 2002b), which is far below the standard replacement fertility rate. This has been attributed to the impact of HIV/AIDS (see Chapter 4, Section 6) and the decreasing fertility rate. Population distribution is rather skewed. The Soweto Township alone holds a very large proportion of the city's population. Rough estimates for the 1990's varied between 1 and 6 million (Musiker and Musiker 2000), but the more recent 2001 Population Census data for Greater Johannesburg estimate that Soweto's population could be between 1.5 and 2.0 million people (Statistics South Africa 2001).

Johannesburg is one of the world's most cosmopolitan cities. The gold rush of the late nineteenth century drew people from all the ethnic groups of the sub-continent, as well as Europeans, American and Australians. As the city grew, traders and entrepreneurs flocked in from India, China, Japan and Eastern Europe. In recent years, many West, East and Central Africans have descended on the city to seek a better life in Johannesburg, putting their own distinctive stamp on the city's profile. South Africa has eleven official languages, with Zulu and Xhosa as the most common mother tongues. English and Dutch-like Afrikaans, are the most widely used across racial lines. In the official context, English tends to be the most commonly used language.

The racist ideology, which shaped the politics of the country until the power shift of 1994, divided the population into four main groups, African/Black, Colored, Indian and White. The city is still characterized by strong ethnic segregation a little more than one decade after the apartheid policy was abandoned (see Chapter 4, Sections 2.2 and 4.1). Even though ethnic mixing has started – for example through Africans moving to the inner city and the lush, well maintained northern suburbs with better access to the city's amenities – most people have remained in highly segregated areas and townships. For example, the dense and deteriorated African areas of Soweto, Alexandra in north-east and Orange Farm in the south still have close to 100% Africans. There are also areas with about 90% Indians (Lenasia) or Coloreds (Eldorado Park) as well as 70–80% Whites (e.g. Roodepoort North, Randburg and Sandton; Statistics South Africa 2003).

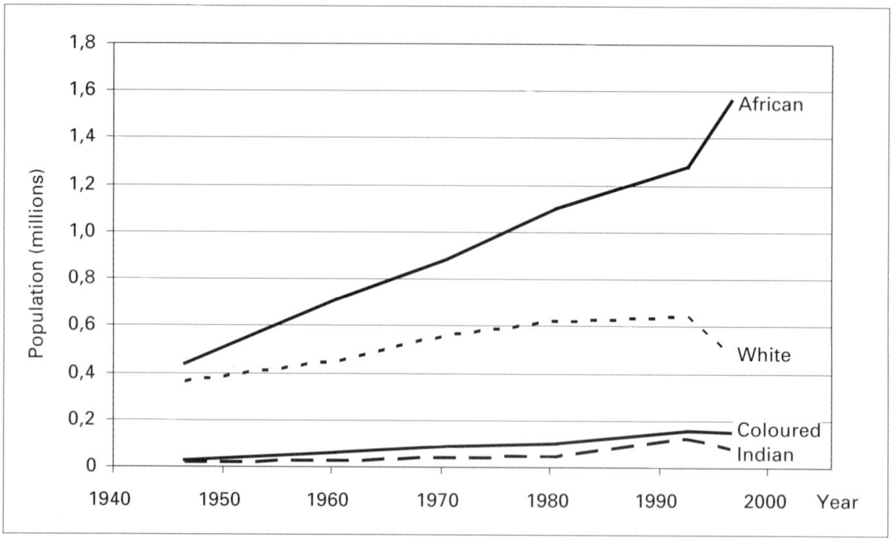

Figure 3-7. Ethnic composition of the population of Johannesburg (Sources: Beall 2002, Republic of South Africa 2003)

2.3 Transforming Economy of Gauteng Hub

According to Beall (2002) the economic growth of Johannesburg can be divided into two contrasting periods and modes of capitalist growth. The first period, which has been characterized as a period of "racial Fordism" (Gelb 1991), lasted roughly from the end of World War II to the onset of the world oil crisis in the mid-1970s. The recovery of Europe created a global demand for manufacturing industry products, which triggered rapid employment growth in Johannesburg. This was a prosperous phase for Johannesburg and the country as a whole, characterized by relatively high rates of growth in employment and in the Gross Domestic Product (GDP). By the middle of the 20th century, Johannesburg's economy relied on manufacturing rather than gold production as the mainstay of the economy.

The second period, which can be characterized broadly as a "post-Fordist" growth, covers the period from mid-1970's to the present. This was a period of decline not only for Johannesburg but also for the entire country as a whole. Unlike many other developed countries, there was a dramatic decline in manufacturing while the tertiary industry progressed slower than for the high-income countries. Regardless, Johannesburg is still a leading industrial and economic city of South Africa and of the African continent (Beall 2002).

With a GDP per capita of about US$ 9,400 (2001 estimates, CIA 2002), South Africa is at the top among African countries and close to the upper quarter of all countries in the world. Being the world's largest producer of platinum, gold and chromium, the mining industry has been the main motor behind the growth during the 20th century, with Johannesburg at its center. Today, South Africa has a well-developed tertiary economy, most of which is centered around the Gauteng conurbation (Johannesburg, Pretoria, Ekurhuleni and other smaller municipalities). Almost half of the employees in the formal sector work in the service sector, which contributes up to two thirds of the GDP. The primary sector represents only 3% of the GDP, even though it is still the main occupation for a large proportion of the population.

Johannesburg is seen as the economic hub of Sub-Saharan Africa. The city has been recognized for being at the forefront of economic activity on the continent. It generates more than 35% of South Africa's *Gross Domestic Product* (GDP) and 10% of the *Gross Geographic Product* (GGP) of the *Southern African Development Community* (SADC). It is also the centre of transportation and shopping in the SADC region. The city hosts 74% of the national corporate headquarters and a network of investment and support systems covering almost the entire continent (Centre for Development and Enterprise 2002). The annual GGP growth rate during the 1990s has been about 2%, which is only slightly above the national growth rate and very low on the international scale. One possible explanation for the low growth rate is that Johannesburg does not include its informal sector in the GGP. The informal sector has grown significantly in recent times. In 1999, it was estimated to contribute 16% of the total employment, mainly in the retail trade sector. But there are other factors affecting the economic growth of the city. Low literacy levels and skill levels, high crime rates and inadequate infrastructure are highlighted as obstacles to growth (City of Johannesburg 2000b).

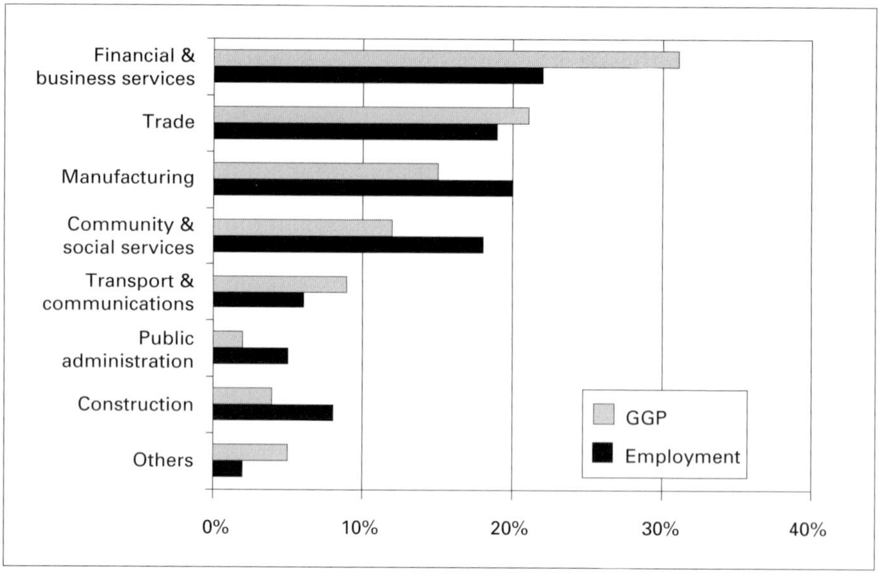

Figure 3-8. Economic sectors in Johannesburg, GGP compared to employment (Source: City of Johannesburg 2002b)

Employment and income – A public survey done for the City Council of Johannesburg showed that job creation was regarded as the first priority problem (City of Johannesburg 2002b). The unemployment level is about 30% and has been increasing over the last few years, in particular due to on-going retrenchments from the secondary and primary sectors as well as the poor matching of skills and opportunities in the tertiary sector. Income distribution in South Africa is extremely unequal. Today, over 50% of the population still lives below the poverty line. The Gini coefficient is almost 60%, which internationally is very high. Only three other countries in the world have a worse distribution compared to that of South Africa (CIA 2002). This aspect is significantly magnified within Johannesburg as the key hub for the generation of wealth and economic growth of the region and the country.

Public finances – Government revenue in South Africa is mainly based on taxes and duties. One third comes from personal income taxes and a quarter from the value added tax (VAT). Education takes the largest share (about 20%) of expenditures followed by social security and welfare (15%) and health (12%). Interest from loans, mainly domestic, stand for as much as 15% of all expenditures (South Africa National Treasury 2003). Most services and development investment are paid through the provincial and local governments. The Province of Gauteng has a budget of about R 23 billion

(US$ 1.8 billion). 95% of the revenue is from national allocation, but there is some revenue from gambling tax and vehicle licenses. More than 80% of the provincial spending goes to education, health and social services. Different development projects like the Alexandra renewal and the *Gautrain* were budgeted to R 2.2 billion for the financial year 2002–2003. The city of Johannesburg has a budget of around R 11 billion (US$ 0.8 billion) based mainly on service charges. The budget can be divided into two blocks. Approximately 60% accounts for the core administration and the rest of the twelve *Utilities, Agencies and Corporate Agencies* (UACs). The core administration is financed mainly with property rates, regional service levies and other interests and charges. This finances the work at the municipal departments and regions. Some of the UACs are financed through their own charges, even if many are dependent on council subsidies or grants, which have limited opportunities of charging full service costs. This is expected to increase over the coming years (City of Johannesburg 2002a).

Economic planning – The economy of Johannesburg strongly reflects successive waves of development and decline, which have seen the city move away from mining and industry production towards an economy fundamentally based on services and trade as well as some high value manufacturing. This trend is in line with global trends and the city's comparative advantages. While crime and unemployment are seen as major stumbling blocks for economic growth in the country, the city is promoting small, micro and medium enterprises (SMMEs) as strategies to aid economic development in the city, reduce unemployment and help the citizens of Johannesburg become more market-connected. Johannesburg has a rich history of developing and implementing sophisticated and advanced programs, which deliver "economies of agglomeration." Unfortunately these economies of localization are in place for heavy, primary production sectors, most especially gold mining, iron and steel industries. As these industries decline and their contribution to the GGP and employment growth decreases, the economies of localization present in the city become more outdated (City of Johannesburg 2002a).

3. GABORONE – AN EMERGING URBAN OASIS

3.1 Origins of the Gaborone Urban Village

The knowledge of contemporary urban developments in Botswana can be better acquired if we distinguish some important factors. According to Silitshena and McLeod (1998), there are three reasons behind the formation of

Botswana's modern towns: (i) the need to exploit mineral deposits, (ii) the development of commerce, and (iii) the need to create a new administrative capital. Gaborone was created as the capital of the new state of Botswana as the time of independence approached. Gaborone was built to satisfy two essential needs: the effective administration of Botswana and the promotion of a modern economy (Silitshena and McLeod 1998). Compared to other emerging African capitals, Gaborone is a relatively young city (only 41 years old), but is one of the fastest growing cities on the African continent (Mosha 1996).

Today, it is not difficult to trace the short historical genesis of Gaborone since the times of the Bechuanaland protectorate, which was part of the Cape Colony known as the "Imperial Reserve." During this period, the British governed the Bechuanaland Protectorate from Mafeking, which is today on the territory of South Africa. Gaborone's ultimate location was chosen and fixed through the initiatives of British and tribal authorities at that time. In the search for an adequate location for a railway station (later known as Railway Siding) and a controlled crossroad and potential site for a police headquarters, a suitable area was recognized just west of the Ngotwane River at a place then known as Lokgolobotlo (today Tlokweng).

In 1890, a small fortification was constructed at the Chief Gaborone's Village (known as Fort Gaberones), and was followed by police housing, stables for their horses, and a prison. This was the start of Gaborone as a settlement, then known as Gaberones Village. Ambitions, however, did not stop there. As developments slowly proceeded, Gaberones grew. Several landmarks from that period deserve particular attention. The Village became the *Prison Service's Headquarters*, which moved the police up from Mafeking, a veterinary office followed, the *Public Works Department* was established mainly to build houses and improve and maintain the earth roads, a sports club (the Gaberones Club) was formed and a trading store was opened (Campbell and Main 2003).

The ascendance of Gaborone to its present position as the prime urban centre of the republic of Botswana began in 1961, when the *Bechuanaland Protectorate Legislative Assembly* recommended to Britain that a capital town should be built at Gaberones, although the British Government expressed the view that Francistown would be a better place. However, the site occupied by present day Gaborone had several advantages: it was straightforward to build a dam in the area, the site lay on Crown Land and had several local administrative offices and it was accessible to six of eight major tribes (Silitshena and McLeod 1998). Gaborone assumed the status of the nation's capital when Botswana attained political independence in 1966.

The shaping of the new town began with an original plan for 20,000 people and spatial provisions for a National Assembly, Government Offices, power station, hospital and schools. The original plan was designed by the *Public Works Department* from Mafeking, in 1962. The project for a contemporary new town was accomplished in 18 months. With the construction of the Government ring and the rim of first settlements, Gaborone started to be shaped on the principles of a Garden City and functionalistic models.

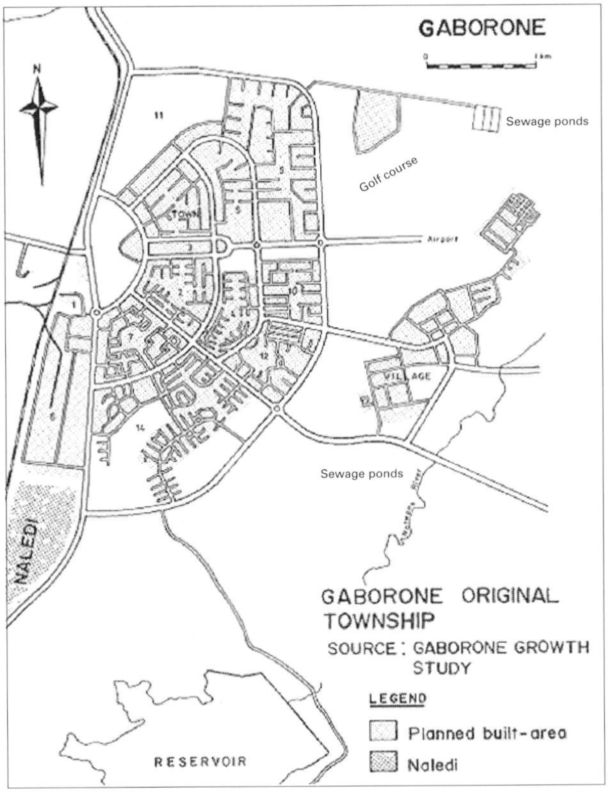

Figure 3-9. Gaborone in 1963

The *Protectorate Administrative Headquarters* moved piecemeal from Mafeking in February 1965, before the completion of the new town. Soon after independence, its name was changed to Gaborone and in 1986 it attained "city" status. After the capital moved from Mafeking to Gaborone, the town rapidly expanded, requiring the expropriation of two neighboring farms, Broadhurst and Bonnington (Campbell and Main 2003).

From a modest small town with 3,855 residents the population of Gaborone reached nearly 200,000 people in 2004. Today, Gaborone City covers some 19.6 km^2. Against the background of a small national population with a widely acknowledged slow growth rate, the events in Gaborone City are truly evolutionary. No doubt, this presents a major challenge for government officials and planners.

3.2 Changing Desert Environment

Gaborone is the center of an area packed with sites of natural, archaeological and historical interest. However, today's Gaborone environmental stability is in question due to the radical change (see Chapter 4, Section 8.3). In less than a century, the large-scale intervention – in contrast to the very sensitive and fragile natural environmental setting – brought to light a new image of Gaborone's small urban enclave and its surroundings: The natural and built-up environments seem to be melting together as spatial structures in the heart of a tremulous "green garden." Gaborone and its present inhabitants have rediscovered the previously forgotten pages of natural and human history by piecing together fragments of memories from pre-colonial, colonial and modern times. The city's expansion to an urban agglomeration is evitable. Though Gaborone is built on a relatively small scale compared to other African capitals, over several decades it has faced a litany of environmental problems. These can be easily distinguished as one moves through the city's districts, neighborhoods and its sub-urban periphery of Tlokweng, Mogoditsane, Gabane, Metsemothlaba and Mmopane.

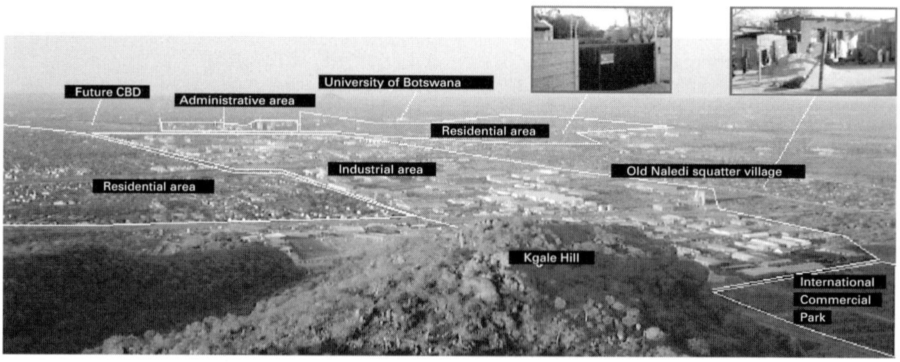

Figure 3-10. View of Gaborone from Kgale Hill (Photos: M. Keiner)

3.3 New Town and Garden City Heritage

The population and economic growth have had a direct impact on the spatial development of the city. In 1963, a Gaborone Master Plan was prepared with the intention of developing Gaborone into a garden city. Essentially, this plan proposed a comparatively low-density form based on a neighborhood model with generous space provisions for pedestrian movement. Subsequently, the city has continued to grow within a low densification framework. For years, one-storey buildings and large plots reflected the nature of the Gaborone city skyline. The lack of land densification, infilling and vertical expansion led to an urban sprawl.

The continuous underestimation of population growth in several preceding planning documents and studies brought a great number of consequences, like shortages in serviced land for housing as well as stress on the existing infrastructure and other facilities.

For example, aerial photos of 1966 (Figure 3-11) and 1998 (Figure 3-12) show the rapid change of land use patterns in the center of Gaborone. Gaborone is becoming a city spreading over a wasteland area and this process cannot be described as sustainable. The enormous land consumption has various adverse effects. Traveling distances between places of living, shopping and working become long, resulting in frequent traffic jams. Residential and commercial developments are built on fertile land, sometimes in areas prone to flooding or on sensitive historical locations without assessing previous land uses (e.g. River Walk near a former landfill). Developing costs for land preparation and infrastructure are increasing.

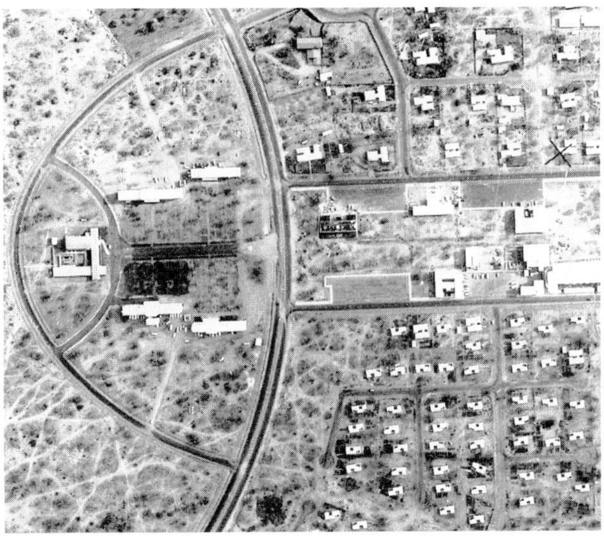

Figure 3-11. Aerial photo of Gaborone 1966
(Source: Government of Botswana 2000)

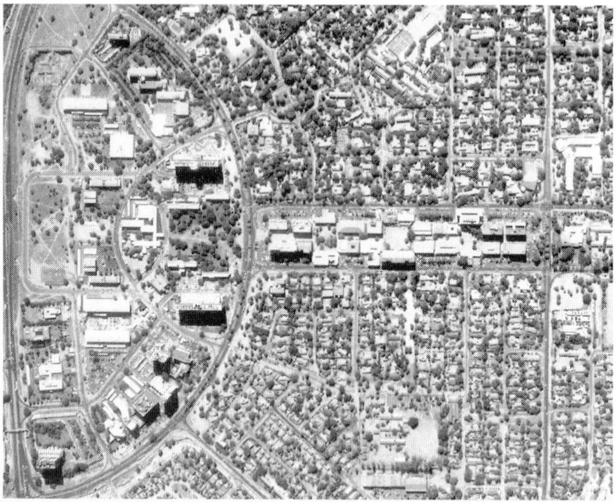

Figure 3-12. Aerial photo of Gaborone 1998
(Source: Government of Botswana 2000)

Today the city center with adjacent industrial and commercial districts is surrounded by sprawling residential enclaves. Some areas are still unplanned, such as Old Naledi where a squatter settlement is being developed. Furthermore, the ongoing uncontrolled and explosive expansion has led Gaborone to overspill into the perimeters of the urban frontier. A first remedy to the mushrooming of Gaborone was made during the 1970's when the first acquisition of surrounding private freehold farms in the north (e.g. Broadhurst farm) and west of the railway line took place, in order to allow further expansion of the capital.

Also neighboring tribal areas were and are still affected, as the satellite settlements around Gaborone have been growing at annual rates of 16% and more (Molebatsi 1996). One important result of the urban sprawl was a loss of arable land, seriously impacting the urban fringes of Tlokweng, Mogoditsane, Mmopane, Gabane and Metsemothlabe. In 1981, more than 80% of the population of the Gaborone Area lived in the city. The surrounding villages were small and experienced only modest growth. Twenty years later, the percentage of Gaborone Area inhabitants who live in the city has dropped to less than 70%. In the two largest dormitory villages Tlokweng and Mogoditshane, the number of residents almost doubled between 1991 and 2001. The growth rate of Gaborone City is now much lower than that of the surrounding area. Hence, the Gaborone Area is becoming an agglomeration where people mainly live in the suburbs and work in the central city, a development supported by existing transportation practices (public or private transport). Commuter traffic, the concentration of the economy and cultural activities on the central city together with related phenomena, are challenges that confront these areas.

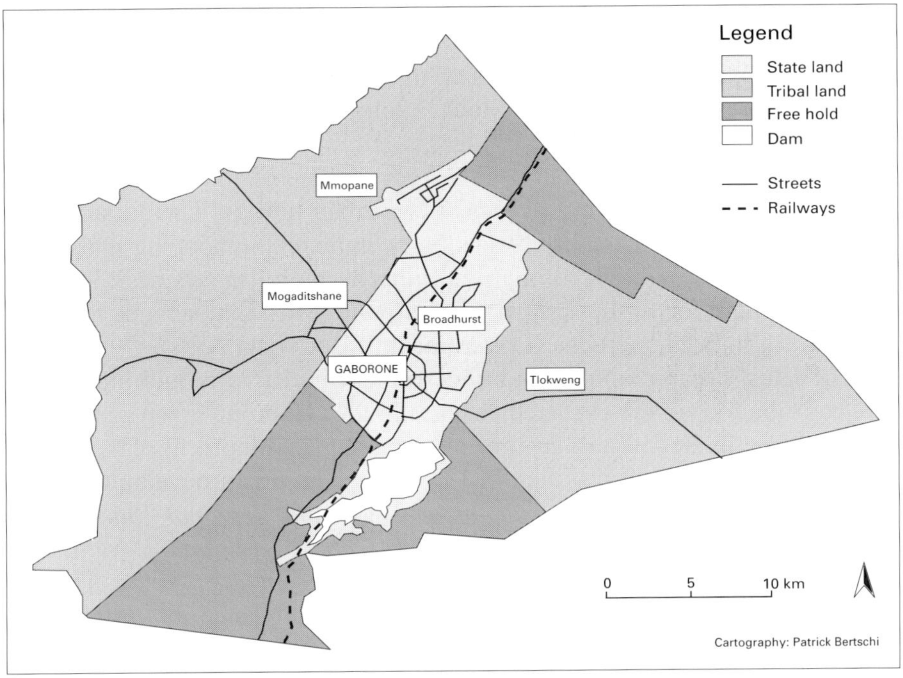

Figure 3-13. Land tenure in the Greater Gaborone Region

Today, it is clearly visible that the urban form of the city of Gaborone is a combination of concentric, linear and gridiron patterns. The older parts of the city have concentric and cul-de-sac road shapes whereas in the new blocks in Gaborone West, the linear and gridiron morphology is becoming predominant. The city is composed of a series of sub-neighborhood modules consisting of 20–30 plots clustered around major open spaces, primary schools and civic and community plots supported by a well functioning secondary road network. These subunits are again clustered together in the form of neighborhoods and super-blocks delineated by primary roads and consisting of local centers, schools, commercial and industrial outlets, parks and open spaces. The sociology of urban layouts confirms the presence of various and mixed socioeconomic and cultural poles, where an average Gaborone inhabitant is still not able to identify himself with the streets and neighborhood where he lives and also with a considerably larger territorial unit such as Tsolofelo, Broadhurst, Block 5, Gaborone Village and others.

Though Gaborone is built on a small scale, as the capital city, it seats Government ministries, the National Assembly, the House of Chiefs and the

National Archives. Each one can be distinguished as one moves through an attractive park called the Government Enclave. At the same time, Gaborone is the centre of one of the most successful economies in Africa. Many representative buildings and public areas provide a variety of attractive landmarks for local urban dwellers and visitors. These are the Mall, Gaborone Civic Centre, the Pula Arch, the University of Botswana, the National Museum and Art Gallery, the Botswana Stock Exchange, the National Sports Stadium, and Gaborone's golf, cricket, tennis and rugby clubs and many more.

Rapid expansion has taken place in every sphere of community life, with high-rise buildings towering over neat, suburban houses. There has been a noticeable demand for quality residential, commercial (shopping malls) and industrial properties in Gaborone, with improvements in business activities indicating room for further property investment. Most major roads in Gaborone have also been upgraded (Government of Botswana 2003).

The latest urban planning and development initiatives could make positive contributions to the expected changes in the Gaborone urban profile. For the first time in Botswana's history of economic development planning, the Gaborone City Council was required to prepare an urban development plan in conjunction with the *Department of Town and Regional Planning.* In 1997, these two institutions prepared the *Gaborone Urban Development Plan I: 1997–2003 Towards Sustainable Economic Diversification.* The aim of this newly approved document was to provide guidance for coordinating development programs in government, non-government, parastatal and private sectors.

This document expresses in clear terms the policies and objectives that will guide and shape the future of the City of Gaborone with the implementation of the *National Development Plan 8*. The plan identifies how scarce resources can be used to satisfy wants and needs. The plan is said to be community-driven as it embodies community needs, aspirations and goals. The plan covers a wide spectrum of diverse, but integrated issues, such as environmental issues, population characteristics, settlements, housing, infrastructure, commercial and industrial development, community services and tourism. The Plan is thus the first of its kind prepared for the city to integrate physical and socioeconomic issues in an urban planning initiative, while previous plans were concerned only with the physical element. The *Gaborone Urban Development Plan I* therefore declares a different approach to future urban planning of Botswana's capital.

3.4 Local Democracy

The people of Botswana are known for their multiparty democracy and long cherished public consultation institutions. Development issues are discussed in various fora starting from the grassroots to the parliament. The smallest of these is the "kgotla" (see Chapter 7, Section 1.7), which is a local institution in which community members can articulate their problems and find solutions to development problems.

Physical and economic plans are prepared through a system of public consultation and participation processes before they can be adopted and implemented. The public is therefore given a chance to participate in project formulation, evaluation and implementation. The development planning process start with the *Gaborone City Council* (GCC) holding meetings in the communities through their "kgotla" and ward committees to come up with proposed development projects. Ideas gathered from these meetings are then used to draw up a development plan. The development plan draft is then revised through workshops with all stakeholders (government departments, municipal departments, parastatals, private sector and the general public). Once approved, a final plan is then prepared for implementation. This way, the plan development process involves all stakeholders and is accepted and effective, although subject to criticism from time to time.

3.5 Economy of Diversification

At the time of political independence in 1966, Botswana was one of the poorest countries in the world. The nation's budget was mainly financed by grants from abroad, particularly from Britain. By then, the major contributor to the GDP was the agricultural sector, accounting for over 50%, and beef was the leading foreign exchange earner (Salkin et al. 1997). Immediately after independence valuable minerals were discovered, including diamonds, copper, nickel and coal. As mining activities started, Botswana became the fastest growing economy of the continent, and in 1980, diamonds had well overtaken beef as the country's leading foreign exchange earner. Today, services and construction sectors are the fastest growing segments in Botswana's economy but even so, diamond mining is still predominant, accounting for over a third of the GDP. Still, the rather prosperous economy of Botswana is not matched by low rates of unemployment (the estimations vary between 15.8%, 21.5% and 40%; Government of Botswana 2000, UN-Habitat 1998 and CIA 2003 respectively). Since mining isn't a long-term revenue, government is concerned about the sustainability of its economy.

Gaborone plays a significant role in the nation's development. As the seat of the government as well as centre of major industries, the city propels the economy of the country. Unfortunately, there are only a few manufacturing industries located in the city as investors find it difficult to invest here due to red tape in company registration as well as the country's small population. As a result, manufacturing, which is the engine of industrialization, has remained weak in the city. The growth sub-sector has not been able to match the rate of urbanization and open unemployment caused by a weak manufacturing base must be mitigated by the engagement of the informal sector.

An increase in the number of people in the informal sector of the city's economy has been observed. The activities in this sector, which include food vending, the sale of vegetables and fruits, welding and carpentry etc, utilize city space not earmarked for such activities. These activities are strategically located where they can attract customers along busy thoroughfares, in close proximity to circles, and in designated open spaces or squares. Most of the structures they operate in are constructed with temporary materials like cardboards, corrugated sheets, discarded cartons, etc. Apart from the eminent visual pollution of these structures, they also pose the danger of converting the cityscape into a slum. This situation lends itself to a consideration for the development of open markets and market stalls in the city. Although these have not been built yet, the informal sector continues to be the second major employer in the city, currently standing at 35%. Government is the main formal employer, accounting for 47% of jobs including parastatals (Mosha 1999).

On the whole, it can be said that Gaborone City has the best potential of all settlements for becoming the engine or prosperous, productive and innovative centre in the country. This is bound to result in the diversification of the economy, which will intensify production in the non-mineral sector and indeed provide full employment so that job opportunities are in balance with the number of job seekers.

The political stability, steady economic growth, liberal economic exchange control regime coupled with the fact that the country is one of the fastest growing emerging markets in the African continent, make the city of Gaborone which is the focal point of all these activities and processes highly conducive for the setting up of a Financial Services Centre for the *South African Development Community* (SADC) sub region. Gaborone is well connected by rail, road and air to other capital cities in SADC. For the city and the government to continue to find ways to simulate economic growth in the private sector, both medium and long term, there should be adequate provision of public infrastructure, e.g. communications and roads, finance etc. This, together with human capital development, will lead to improvements in productivity. The Gaborone City Council supports the view that a dynamic private sector, which is essential sustaining a balanced development of the economy, is encouraged to grow. It believes that the best way to solve unemployment is to have a vibrant small and medium business community, but this can only come about if commercial banks and others come to realize that they can do good business with small business entrepreneurs.

NOTES

1. After the first paragraph, this section is reproduced, with permission and slight update, from Zegras and Gakenheimer (2000).
2. Except for the first and last paragraphs, this section also comes from Zegras and Gakenheimer (2000), with slight updates.

REFERENCES

Allan, K., G. Gotz and C. Joseph 2001. *Johannesburg: An African City in Change. A Recent History of Local Government in Johannesburg, 1995–2001.* Cape Town: Zebra Press.

Barros, A.D., A.Õ. Inostroza., M.A. Allison 2002. *Santiago de Chile*. México: Grijalbo.

Beall, J. (Ed.) 2002. *Uniting a Divided City: Governance and Social Exclusion in Johannesburg*. London: Earthscan Publications.

Beavon, K.S.O. 1997. *Johannesburg: A City and Metropolitan Area in Transformation.* In C. Rakodi (Ed.) *The Urban Challenge in Africa – Growth and Management of its Large Cities.* Tokyo, New York, Paris: UNU Press: 150–191.

Bromley R. and G. Jones 1996. Identifying the Inner City in Latin America. *Geographical Journal*, 162 (2): 179–190.

Campbell A. and M. Main 2003. *Guide to Greater Gaborone*. Gaborone: The Botswana Society, Printing and Publishing Company Botswana.

Carrión D.M. 2001. Democracy and Social Participation in Latin American Cities, *Development in Practice*, Vol. 11, Nos. 2 & 3: 208–216.

Catholic University of Chile 2004. *DIMSUD Synthesis Report: Towards Sustainable Urban Development in Santiago, Chile.* Unpublished working paper.

Central Bank of Chile 1999. *Statistical Synthesis of Chile 1986–1996*. Santiago.

Central Bank of Chile 2000. *Statistical Synthesis of Chile 1996–2000*. Santiago.

Central Bank of Chile 2003. *Statistical Synthesis of Chile 1998–2002*. Santiago.

Centre for Development and Enterprise (CDE) 2002. *Johannesburg: Africa's World City – A Challenge to Action.* Johannesburg.

CIA 2002. *The World Fact Book 2002 – South Africa.* Central Intelligence Agency, United States Government. http://www.cia.gov/cia/publications/factbook/geos/sf.html, retrieved on April 4, 2003.

CIA 2003. *The World Fact Book 2003 – Botswana.* Central Intelligence Agency, United States Government. http://www.cia.gov/cia/publications/factbook/geos/bc.html, retrieved on March 12, 2003.

City of Johannesburg 2000b. *State of the environment.* http://ceroi.net/reports/johannesburg/csoe/Default.htm, retrieved on March 3, 2004.

City of Johannesburg 2001a. *Johannesburg – An African City in Change*. Johannesburg: Zebra Press.

City of Johannesburg 2001b. *Spatial Development Framework*. Department of Development Planning, Transportation and Environment, City of Johannesburg.

City of Johannesburg 2002a. *Integrated Development Plan 2002/03*.

City of Johannesburg 2002b. *Joburg 2030 Full Report.* Corporate Planning Unit, City of Johannesburg.

City of Johannesburg, 2002c. *Local Integrated Development Plan. Region 2*. Development Planning and Facilitation, City of Johannesburg.

Crowley, W.K. 1995. Order and Disorder – A Model of Latin American Urban Land Use. In Turbeville III, D.E. (Ed.) *Yearbook of the Association of Pacific Coast Geographers*. Corvallis: Oregon State University Press: 9–31.

De Mattos, C.A. 1996. Avances de la globalización y nueva dinámica metropolitana: Santiago de Chile, 1975–1995. *Eure,* XXII: 39–60.

Errázuriz, A.M. 1987. *Manual de Geografía de Chile*. Santiago: Editorial Andrés Bello.

Ford, L. 1996. A New and Improved Model of Latin American City Structure. *Geographical Review* 86 (3): 437–440.

Galafassi, G.P. 2002. Ecological Crisis, Poverty and Urban Development in Latin America. *Democracy and Nature*, Vol. 8, No. 1: 117–129.

Gelb, S. 1991. South Africa's Economic Crisis: An Overview. In S. Gelb (Ed.) *South Africa's Economic Crisis.* Cape Town: David Philip: 1–32.

Gobierno Regional Metropolitano (GoRe) 1995. *Estrategia de Desarrollo: 1995–2000*. Santiago.

Government of Botswana 2000. *Botswana National Atlas*. Gaborone.

Government of Botswana, 2003. *Botswana Review, 23rd Edition*. Gaborone: B & T Directories.

Griffin, E. and L. Ford 1980. Cities of Latin America. In Brunn, S. and J. Williams (Eds.) *Cities of the World*. New York: Harper and Row: 199–240.

Griffin, E. and L. Ford 1983. A Model of Latin American City Structure. *Geographical Review* 70: 397–422.

Heritage, A. (Ed.) 2003. *Financial Times World Desk Reference*. New York: Dorling Kindersley.

Heritage Foundation 2004. *Index of Economic Freedom*. http://www.heritage.org/research/features/index/, retrieved on May 6, 2004.

Huchzermeyer, M. 2001. Housing for the Poor? Negotiated Housing Policy in South Africa. *Habitat International.* (25): 303–331.

IMC 2003. *SouthAfrica.info – the Official Gateway*. The International Marketing Council of South Africa. http://www.safrica.info/, retrieved on May 10, 2003.

Instituto Nacional de Estadísticas (INE) 2002. *Chile: Ciudades, Pueblos y Aldeas (Censo 1992)*. Santiago.

Mahanyele, S. 2002. *Towards Sustainable Built Environment*. Minister of Housing's keynote address at pre-WSSD conference. University of the Witwatersrand. Johannesburg. August 2002.

Molebatsi, C. 1996. Towards a Sustainable City: Gaborone, Botswana. *Ambio* Vol. 39, No. 4: 205–210.

Mosha, A.C. 1996. The City of Gaborone, Botswana – Planning and Management. *Ambio* Vol. 25, No. 2, March 1996. Stockholm: Royal Swedish Academy of Sciences: 118–125.

Mosha, A.C. 1999. *Urban Poverty, the Case of Gaborone, Botswana.* A UNCHS/Habitat/Ford Foundation Report. Gaborone.

Mosha, A.C. 2002. *Local Democracy and Decentralization in Botswana*. A UNCHS Report. Gaborone.

Musiker, N. and R. Musiker 2000. *A Concise Historical Dictionary of Greater Johannesburg*. Cape Town: Francolin Publishers.

Pacione, M. 2001. *Urban Geography – A Global Perspective*. London: Routledge.

Potter, R.B. and S. Lloyd-Evans 1998. *The City in the Developing World.* Harlow: Addison Wesley Longman.

Redclift, M.R. 1973. Squatter Settlements in Latin American Cities: Response from Government. *Journal of Development Studies* Vol. 10(1): 92–109.

Republic of South Africa (RSA) 1994. *The Reconstruction and Development Program – A policy framework.* http://www.polity.org.za/html/govdocs/rdp/rdpall.html, retrieved on March 26, 2004.

Republic of South Africa, Municipal Demarcation Board 2003. *Municipal Profiles – City of Johannesburg.* http://www.demarcation.org.za/Munic_Profiles/New_layout/index.html, retrieved on April 4, 2003.

Salkin, J.S., D. Mpabanga, D. Cowan, J. Selwe and M. Wright 1997. *Aspects of the Botswana Economy. Selected papers.* Gaborone: Lentswe La Lesedi (Pty) Ltd.

Silitshena, R.M.K. and G. McLeod 1998. *Botswana – A Physical, Social and Economic Geography.* Gaborone: Longman Botswana.

South Africa National Treasury 2003. *Budget 2003.* http://www.treasury.gov.za/documents/budget/2003/ene/default.htm, retrieved on March 26, 2004.

Statistics South Africa 2001. *Census 2001.* http://www.gov.za/reports/2003/census01_key.pdf, retrieved on April 4, 2003.

Statistics South Africa 2003. *Census 1996 – City of Johannesburg.* http://www.statssa.gov.za./SpecialProjects/Herman/wards/catb/Johannesburg.htm, retrieved on April 4, 2003.

Verhasselt Puppinck, D. 1997. *Chile.* Santiago: Editorial Kactus-Sipimex.

Violich, F. and R. Daughters 1987. *Urban Planning for Latin America.* Boston: Oelgeschlager, Gunn & Hain.

UNESCO 1998. *World Science Report 1998.* Paris.

United Nations Human Settlements Programme (UN-Habitat) 1998. *Global Urban Indicators Database 2 (1998 data).* http://www.unchs.org/programmes/guo/documents/1998.zip, retrieved on November 20, 2002.

Zegras, C. and R. Gakenheimer 2000. *Urban Growth Management for Mobility: The Case of the Santiago, Chile Metropolitan Region.* Prepared for the Lincoln Institute of Land Policy and the Cooperative Mobility Program. Cambridge: MIT.

Chapter 4

CHALLENGES FOR SUSTAINABLE URBAN DEVELOPMENT

Daniel K. Irurah[1], Björn Malbert[2], Pål Castell[2], Jaan-Henrik Kain[2], Branko I. Cavric[3], Aloyse C. Mosha[3]

[1]University of the Witwatersrand, [2]Chalmers University of Technology, [3]University of Botswana

Abstract: This chapter focuses on the key challenges facing cities in developing countries. Observations from the case study cities of Gaborone, Santiago and Johannesburg have indicated an overwhelming range and scale of challenges with extremely limited opportunities. Deterioration of socioeconomic, environmental and institutional capacity has been noted as a recurrent trend in all three cities. Equally, prioritization of socioeconomic and local environmental challenges has meant that global environmental concerns have not been accepted as urgent local issues. Consequently, the expectation that, through sustainable development, developing countries will catapult themselves into cleaner technologies (smaller ecological footprint with increasing standards of living) cannot be justified on the basis of the prevailing trends in the case study cities at this point. It is therefore clear that efforts to address sustainable urban development in a comprehensive manner will require many more resources and several times the capacity currently in place within such cities.

Key words: Sustainable development as process and outcome, urban sprawl, rural-urban migration, inter-urban migration, economic development, urban poverty, inequality, tertiarization, job-less growth, socioeconomic segregation, apartheid, housing, health, HIV/AIDS, woman empowerment, infrastructure, urban transport, environment, ecological footprint, air pollution, governance, public participation, institutional framework, financing

M. Keiner et al. (eds.), From Understanding to Action, 75–114.

1. INTRODUCTION

1.1 Aim and Scope of the Chapter

The challenges for sustainable urban development in the three case cities are following global trends and driving forces, however differ from city to city depending on the various local geographical, institutional, cultural, and historic conditions of each urban situation. The main challenge concerns the shaping of a constructive meeting between global and local driving forces, strategies, and actions for sustainable urban development.

The aim and scope of this chapter is to further review and discuss challenges recognized in the case studies in order to reveal similarities and differences supporting the synthesized conclusions of the DIMSUD research. Thus, information that can be generalized as being valid elsewhere will be separated from what is only valid in the specific situations of the case study cities, or information that still needs to be transformed if used in another urban context. This chapter will also shape the point of departure of Chapter 5, where opportunities and possible strategies for sustainable urban development in the case study cities will be presented and discussed accordingly.

1.2 A Framework of Trends and Trajectories

The vision of *sustainable development* is presented and discussed in Chapter 1. The possible definition and interpretation of this global vision has been intensively debated in the realm of politics as well as research since it was more widely launched by the Brundtland commission (WCED 1987). Without dwelling on the detailed appraisal of such debates, it is important to conclude here that sustainable development is a newly emergent system of thinking about humanity, development and the natural resource-base in a holistic and interactive manner. What is new about this system of thought is the recognition of the "agency" role of the natural environment in relation to development as opposed to the passive role predominantly assumed in conventional theories of economics and development.

Under this conceptualization, sustainable development can be viewed simultaneously as a process and an outcome. As a process, sustainable development envisions humanity's transition from a less favorable level of quality of life to a more favorable one. When such transitions are mapped over a series of time periods, sustainable development would be seen as a trajectory with consistent improvements in levels of sustainability. Such an interpretation can be defined best through scenarios and their related trajectories (see

Figure 4-1). The following are three scenarios that would help clarify the concept:
- Worsening trend scenario
- Status quo scenario
- Improving trend scenario

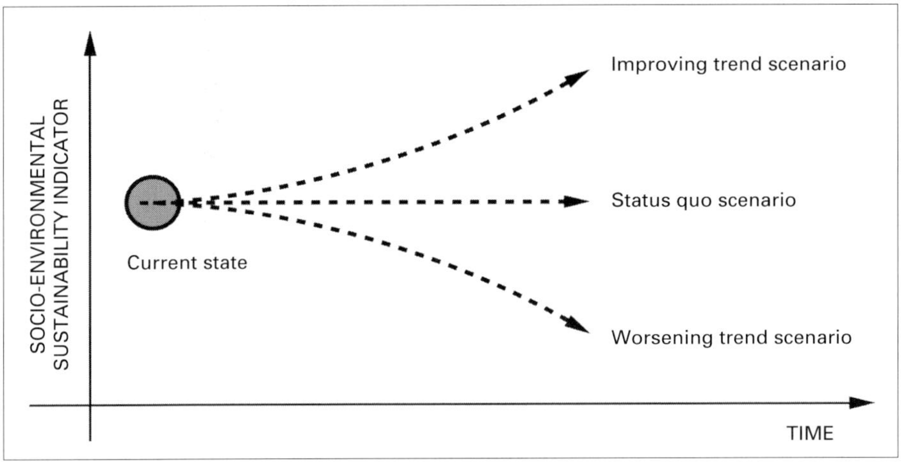

Figure 4-1. Trends and trajectories as a framework for describing challenges and opportunities

The first one represents the "do-nothing" option, where humanity opts not (or is unable) to intervene with regard to increasing awareness of the negative socio-environmental impacts of economic growth and development. Consequently, the socio-environmental state risks to decline, thus threatening the current systems as well as constraining the possibilities for future generations. The second scenario represents action scaled to ensure no further deterioration of the socio-environmental state due to development impacts. The last scenario represents action scaled at a level that ensures improvements to the prevailing state of socio-environmental systems.

Although the above interpretation provides a simple approach towards defining and communicating the concept of sustainable development, it poses a major challenge with regard to the formulation and application of a socio-environmental sustainability index to monitor for the generation and representation of the trajectories of the different scenarios. This challenge is composed of two parts. Firstly, there is the question of the range of issues/factors to be monitored and analyzed in order to compile the index of negative impacts on socio-environmental systems, which manifest themselves in a variety of forms and levels. Secondly, the relative importance of the selected range of factors and their respective weighting in the derivation of such an index would entail a complex process of stakeholder consultation and consensus building. Since one would also expect such weighting to vary over time based on the relative change of impacts and related stakeholder-perceptions, the validity of the resultant index as a framework for comparison between communities, countries, cities and regions or between one period and another would be significantly weakened. The use of indicators and monitoring systems will be discussed further in Chapter 6.

1.3 Towards a Framework for Challenges and Opportunities

As the main focus of this chapter and the following Chapter 5, the challenges and opportunities of future urban development will be elaborated in terms of a larger framework in order to provide a systematic approach for both. This section aims to provide the working definitions and operationalization of the two concepts in order to facilitate their substantiation in subsequent sections of the two chapters.

A *challenge* is a general term for tasks or problems experienced as difficult and threatening, but at the same time stimulating and inciting. In this case, there are some different kinds of challenges that we highlight. One kind is the elimination of those elements that contradict the recognized ideal of the "good society," such as starvation, diseases, homelessness, poverty and other suffering. Another kind includes *trends and pressures* that threaten to create severe problems in the future or worsen existing problems, such as rural-urban migration, increasing car traffic, declining economy, environmental degradation etc. A third kind of challenges raised here involve circumstances that hinder the possibilities to create a positive development, such as limited funding, lack of governmental capacity, lack of knowledge etc.

What is a challenge in one situation does not have to be a challenge in another situation or in another place. For example, a high rate of urbanization does not need to be a major threat in a context where institutional capacity (policies/governance, human and financial resources) exists to meet the demands and needs arising from urbanization. It is in this context that the newly developed countries of East Asia (as *Newly Industrialized Countries* – NICs, often referred to as the "Asian tigers") have managed to achieve higher standards of living for a majority of their populations during periods of high rates of urbanization. It is also in this context that most countries in Sub-Saharan Africa and in Latin America have experienced stagnation or declining standards of living related to high rates of urbanization (Rakodi 1997a: 52, 66; Gelderblom and Kok 1994).

As a result, tackling the manifestations of the mismatch between ongoing urbanization and the capacity to address the resultant developmental demand may lead to a number of sub-challenges (poverty, slums and illegal settlements, absence of municipal services, low literacy levels, unemployment, etc). Such challenges get even more complex as a result of related vicious cycle tendencies centered on urbanization and poverty.

While this chapter focuses on the *challenges*, the following chapter will discuss the *opportunities* in terms of existing and possible strategies for the city authorities. The concept of opportunity is closely related to that of challenges. An opportunity may be described as a favorable junction of circumstances that appears in relation to a challenging problem. An opportunity, therefore, indicates a window for leverage, which once recognized and synergized can result in major gains in the right direction. It is thus a dynamic to align to, unlike a challenge, which is a pressure that is to be countered or controlled.

Opportunities for sustainable urban development can be sought in many dimensions and domains of society. A great potential is, of course, the visions (see Chapter 5), intentions and decisions of urban actors and citizens based on insights and awareness. The individual understanding and awareness is influenced by the local situation concerning natural, human-made, and institutional resources as well as by relations to the external world. These conditions offer options and limitations for social actions that in turn influence the local situation.

The review of the case study cities will be structured around the following themes based on what has been discussed in the case study reports:

- Urbanization and sprawl
- Economic development, poverty and inequality
- Socioeconomic segregation
- Housing
- Health and social services
- Infrastructure
- Environmental impacts
- Governance
- Financing

The aim is to highlight themes where comparisons can be made between at least two of the cities.

2. URBANIZATION AND SPRAWL

It may be argued that "land" or "open space" constitutes the most basic resource of a city. During periods of high rates of urbanization and rural-urban migration, the consumption of this resource manifests itself not only in terms of the intensification in use of existing urban areas but also in the rapid geographic spread of the city or urban sprawl (see Chapter 1, Section 4).

One of the underlying mechanisms of urban sprawl is related to the pricing of land in urban areas, especially where motorized transport has tended to moderate the price through increased supply, i.e. it brings low-cost land in the peripheries into the urban market. Moreover, alternative land uses, such as agriculture and nature conservation, suffer due to their relative inability to compete in terms of generated land values. The interplay of the above factors and how they influence land values and city form has been intensively investigated in a variety of literature on land rent and theories of land values. Thrall (1987), in particular, provides a detailed mathematical analysis of the related dynamics based on the *Consumption Theory of Land Rent* (CTLR). Among other things, he analyzed the impact of transport infrastructure and related costs on land values and urban form (Thrall 1987).

Uncurbed urban expansion generates a number of dilemmas (see Chapter 1, Section 4). Even in the short-term, urban sprawl constitutes an immediate threat to the supply of city inputs, like food and water, but also mined and quarried raw materials for construction and manufacturing industries. Its more long-term environmental impacts include loss of bio-diversity, increased rainwater run-off and related flooding and water and air pollution due to increased car use among other pollution sources.

Furthermore, the distribution of and access to other amenities such as schools, health clinics, infrastructure, recreation, jobs and markets becomes undermined. Contrary to the idea of cities as settlements for the maximization of exchange and cultural growth – based on proximity of a diversity of functions and activities – extremely low overall population density levels characterize contemporary sprawled urban settlements. At the same time, the spatial segregation of functions has predominated planning and land development throughout the 20th century (Lozano 1990). As a result, inadequate densities undermine the attainment of thresholds for the feasible operation of amenities and public transport modes, thus, reinforcing the dependence on private passenger cars for those who can afford them and spatial segregation for those who cannot afford private means of transport (Lozano 1990).

2.1 Rural-Urban Migration in Gaborone

Today, half of Botswana's population lives in urban areas but in the 1980's only 35% of the inhabitants were urban (Government of Botswana 1997). If the projected population growth for Gaborone becomes reality, an additional 250,000 new residents will live there by year 2021, which is more than the population of today's 200,000 residents (Keiner and Salmerón 2003).

The migration into Gaborone is mainly driven by the expectation to find paid employment in the secondary and tertiary economic sectors (Silitshena 1996). In 1994, the average urban household income was more than five times higher than in rural areas (Government of Botswana 1996). Consequently, a major challenge in Botswana is to moderate or redirect the migration between countryside and urban settlements. Here, one potential evidently lies in improving rural living conditions (see Chapter 7, Section 1.1).

The land-take implications of the rapid expansion of Gaborone – with a demand for plots that continuously outstrips supply – have been twofold. First, the municipal boundaries have been extended several times in order to open up surrounding farmland for urban development. For example in 1970, a major acquisition of large farm portions to the north and west of the city was implemented. Second, neighboring tribal areas and villages are also heavily affected, since a large portion of the growing population has settled in surrounding satellite settlements. One explanation for the enormous growth of these settlements is that on tribal land, every indigenous person gets a plot for free. In fact, these villages are growing even more rapidly than Gaborone City itself (Molebatsi 1996) while serving as dormitories for people working in the city centre. The result of this development is a serious encroachment on arable land supporting the urban settlements on the periphery.

One aspect that further reinforces the Gaborone sprawl situation is that the 1963 Master Plan for the future capital (see Chapter 3, Section 3.3) was designed to meet administrative needs with a maximum of 20,000 inhabitants, which is just a tenth of the actual population of the current metropolitan area. Today, the city is confronted with the problem of land shortage to meet the constantly increasing demand for land. Almost all the land within the city boundaries has been committed and the city is now looking to the peri-urban areas for expansion.

2.2 Post-Apartheid Urbanization in Johannesburg

In some aspects, the urbanization process of Johannesburg is similar to that of Gaborone. It has grown to a three million people city in just slightly more than one hundred years and still attracts large numbers of people from other part of South Africa as well as from other countries (Beavon 1997). In South Africa, poverty and lack of opportunities *push* people to leave their homes in rural areas. In addition, hopes and dreams of education, jobs and wealth tend to *pull* people into Johannesburg. Being a major continental hub (economically and culturally), Johannesburg will continue to attract people from the whole sub-Saharan Africa and elsewhere in the world.

Of course, such migration patterns exert intense pressure on urban land. At present, data of the City Council of Johannesburg shows a deficit of 217,000 housing units and targets to reduce that to 50,000 by 2004 (Thale 2002). However, when there is no accessible formal housing, the poor immigrants will provide for themselves in illegal ways. Consequently, the number of informal settlements in Johannesburg is a problem, not only for the affected people, but also for the city in terms of planning and service provision.

The South African apartheid system has not only resulted in social segregation and inequities (see Section 4 below) but has also left very tangible imprints in the physical urban environment. Like other South African cities, Johannesburg is characterized by typical "match-box" housing – large areas, such as Soweto, located in peripheral townships and with high population densities but filled with rows of identical one-storey houses (City of Johannesburg 2000). In Soweto, the standard plot was set to 40 by 70 feet ($280m^2$) with the house in the centre. The floor area ratio in these areas can be estimated at 0.1–1.25, which is almost ten times lower than that found in European inner cities (Vestbro and Algren 1999). However, the population density is very high, with up to 300 people per hectare (City of Johannesburg 2000b).

Yet another urgent issue is the decay of the inner city of Johannesburg. Many companies and services have already left the city centre in favor of new commercial centers in the northern suburbs of the city. The main driving force for this shift is the existing high crime rate and perceived threats from informal uses such as mini-bus taxis and informal trading in the Central Business District.

2.3 Intra-Urban Migration in Santiago

In terms of urbanization trends, Santiago differs substantially from Gaborone and Johannesburg. Although Santiago has been growing steadily throughout the 20th century, census data from the last decade indicates an emerging stabilization of rural-urban migration patterns. Moreover, the overall population growth rate of the city has also declined in recent years.

Nevertheless, the urban area of Santiago continues to grow but now mainly as a result of intra-urban migration. Here, the main trend is a population loss in areas less than seven kilometers away from the city centre and increasing densities further outwards. A number of mechanisms have coproduced these migration patterns, such as low land prices in the urban periphery (favoring social housing projects), ambitions to escape from a deteriorating city centre and a longing for more spacious surroundings, but also changes in land use – from residential to commercial and service uses – have contributed significantly to processes of urban sprawl (see Figure 4-2).

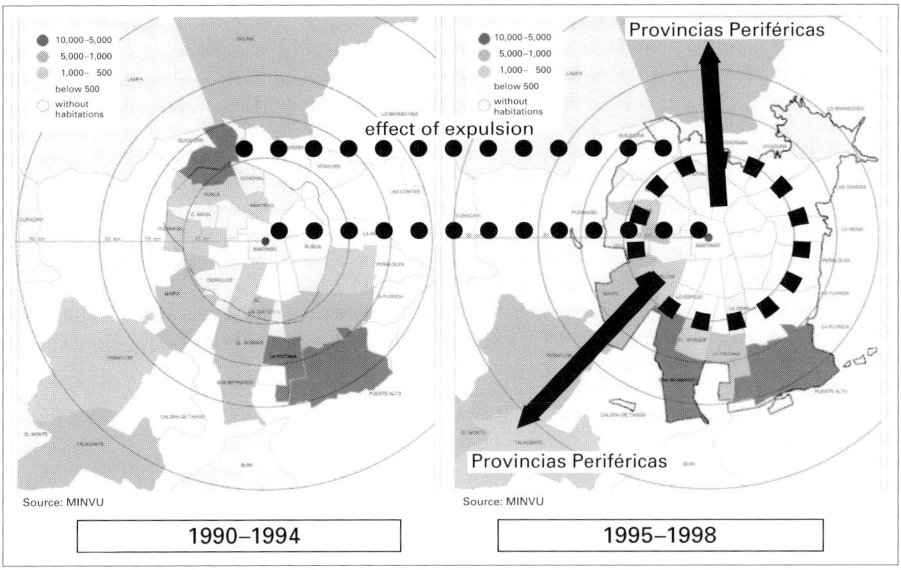

Figure 4-2. Urban sprawl and expulsion of social housing in Santiago (Source: Catholic University of Chile 2004)

The attempt by the central government to manage this development by introducing and maintaining an urban boundary has, so far, not been successful. One reason is that there exist loopholes in current legislation, and also that governmental capacity is limited in relation to existing economic pressures for urban expansion and redevelopment.

3. ECONOMIC DEVELOPMENT, POVERTY AND INEQUALITY

3.1 Local Labor Markets and Private Sector Investment

From a neoliberal market perspective, private sector investment is seen as the engine of economic growth and employment (see Chapter 3, Section 3.5), whereas the role of the public sector is reduced to facilitation through the provision of responsive governance, infrastructure and services. Similarly, it is argued that cities in low, middle as well as high income countries require high levels of externally sourced investments in order to realize and sustain economic growth – economic expansion needed to generate employment opportunities as well as revenues, for example, infrastructure provision.

However, such externally sourced investments mainly aim to optimize returns and therefore seek locations in areas of maximum opportunities but minimal risks and costs. Due to globalization, all cities, therefore, end up competing for these limited investments by enhancing their attractiveness according to their perceived niches of comparative advantage (Sustainable Cities 2000 and 2002). Unfortunately, from this perspective, cities in many low and middle-income countries suffer two major disadvantages (Sustainable Cities 2002). First, in terms of business opportunities, their markets are weak due to high levels of unemployment, low skill levels, low incomes and a weak infrastructure base. Second, in terms of risks to investments, weak governance, unstable political environments, the prevalence of corruption, high levels of crime and threats to property rights are viewed as major threats.

Private-sector strategies to manage these two disadvantages lead to comparatively high establishment and transaction costs and such cities are, therefore, viewed as insufficiently competitive and attractive – especially for footloose global capital (Sustainable Cities 2002). As a consequence, low and middle-income country cities are often trapped in vicious cycles, characterized by stagnation in economic growth, increasing unemployment and poverty, a weak revenue base, the weakening of city competitiveness and attractiveness, and thus further stagnation and so on.

Like most towns in Botswana, Gaborone has its own share of poverty. Estimates indicate that over 40,000 people in the city are now living below the National Poverty Datum Line, and that one third of these people (about 7% of the population) are very poor (Mosha 1999). Both the central government and the Gaborone City Council have initiated various programs to ameliorate this situation. Individuals also have embarked means of addressing the poverty that is affecting them. This situation cannot be left to continue and more sustainable efforts have to be made to address this issue and lessen the burden.

The processes of rural-urban migration discussed in the previous section also further strain labor markets of low and middle-income countries. In the first half of the 20th century – when primary and secondary sectors dominated urban economies – there was a reasonable level of labor absorptiveness as well as congruence between the skill levels of migrating job seekers and the skills required in the labor markets (Sustainable Cities 2002). Since the mid 20th century, however, there has been a major structural shift in most urban and national economies. Today, most economic growth occurs in the tertiary sector (i.e. the service industry) with business and financial services as dominant activities (Sassen 2002).

This structural shift has contributed to two major trends in low and middle-income country cities. Levels of unemployment in the formal sector have been rising, often approaching 50%. Subsequently, levels of employment in the informal sector – self-employment and casual employment in informal/seasonal businesses – have been rising rapidly (Rogerson 1997).

3.2 Tertiarization and "Jobless-Growth" in Southern Africa

The rise in Gaborone's population over the past years has meant an increase in the labor force with little chance of all people getting employment. This is the greatest challenge to the city that has limited resources at its disposal since it is so dependent on the central government. According to the 1991 Population Census, the total labor force in the city was made up of 75,161 workers, of which 44,831 were male and 30,330 were female, with the bulk of these employed either in the formal or informal sectors. The greatest employer was manufacturing and construction at 32% and the government with 31%, followed by sectors like wholesale, retail and communications at 20%, finance, insurance and business services 8% and community and personal services at a 7% and agriculture 1% (Government of Botswana 1991). On the negative side, the economically inactive population numbered 27,563, the bulk of which were women 18,856 as compared to 8,707 men. The majority consisted of housewives followed by students, together forming the group that is prone to poverty and thus needs help. In terms of employment opportunities, the hardest hit age group is 15–19 years at 12% among the city's population. The reason for the high level unemployment for this group is because most of them do not have marketable skills and have attained low levels of education. Females accounted for a 42% unemployment rate with males at 25%. The high youth unemployment rates may be linked to education since many of the Form II leavers – these are scholars having attended school for 10 years and obtaining the *Junior Certificate* in a Junior Secondary School cycle – are hardly employed (Government of Botswana, HIES 1993/94).

Botswana's trend towards a tertiarization of the economy is even more accentuated in South Africa. In present-day Johannesburg, the services sector, and business and financial services in particular, dominate the economic scene both in terms of Gross Geographic Product and employment (see Figure 4-3). This structural transformation is exerting extreme pressure on the city in relation to unemployment. Although the city's economy has been growing at 2% per annum over the decade 1990–2000, job creation only grew at 1.1% per annum (City of Johannesburg and Monitor Company 2000), and currently the unemployment rate in Johannesburg is estimated at 30% (Centre for Development and Enterprise 2002). This trend is often referred to as "jobless-growth", i.e. economic growth that is accompanied by stagnation or even a decrease in employment. The *Joburg 2030* team conducted a household survey in year 2000 to determine the most pressing concerns in the city. Uppermost in the minds of citizens was the need for employment. Fifty percent of people surveyed asked that the Council prioritize job creation. This is not surprising, considering that the unemployment rate in the city has risen to above 30%.

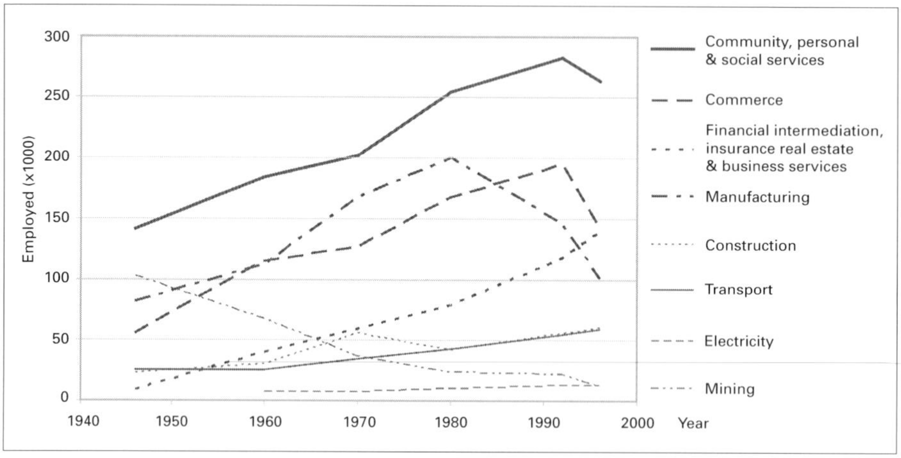

Figure 4-3. Economic sectors in Johannesburg 1946–1996 (Source: Beall 2002b)

It may also be added that the labor-intensive industries of Gaborone suffer greatly from competition by the neighboring South African market, which plays a key role in the whole region. In South Africa, there is a functioning market as well as necessary infrastructure, resources are cheaper and the quality of products is better. As a consequence, local small-scale producers are very much outdone by South African enterprises.

Nevertheless, both in Gaborone and Johannesburg, a major reason for the lack of jobs is the growing mismatch between skill levels of the residents and the skills in demand in a tertiarizing economy. In other words, low levels of literacy and training have limited the capacity of residents to take full advantage of the jobs available in the growing tertiary sector. In addition, the primary and secondary sectors have undergone increasing levels of automation, thus requiring fewer workers but also here with higher skill levels. When combined with a general stagnation in these sectors, the result is extremely low growth, and even decline, in employment. Meanwhile, there has been rapid growth in informal employment levels, leading to a constantly growing trend of the "informalization" of employment (City of Johannesburg and Monitor Company 2000).

However, the structural transformations do not only affect employment rates but also cause an accelerated unequal distribution of income. In informal employment, payment levels are low and job security is poor, whereas the expanding tertiary sector is characterized by substantially higher remunerations. Consequently, although neither Botswana nor South Africa are poor nations, disparities in income levels have been growing rapidly and the income distribution in both countries is extremely unequal (Government of Botswana 2000; CIA 2002). Some estimates claim that as much as half of the populations of both countries are living below the poverty line (see for example UNDP 2003, UN-Habitat 1998, CIA 2002). Moreover, the inequality in income distribution also manifests itself in a gender dimension where a higher proportion of the unemployed and low-level earners, in both formal and informal employment, are women. Consequently, low-income women-headed households are the most vulnerable to poverty (UNDP 2000).

4. SOCIOECONOMIC SEGREGATION

In many developing countries, processes of urban sprawl tend to reinforce socioeconomic segregation and the marginalization of the urban poor. On the one hand, marginalized groups have effectively been kept out of the formal land market system and driven into peripheral, informal and often illegal settlements. These settlements are characterized by extremely high densities, insecure tenure, temporary/informal shelters, inadequate municipal infrastructure and services and extremely low standards of environmental health (see for example Yousry and Atta 1997 for Cairo; Beavon 1997 for Johannesburg; Behrens and Wilkinson 2003 for South Africa in general).

On the other hand, more affluent groups of the population – particularly in developing countries – are forming new urban peripheral enclaves in search for high quality infrastructure provision, services and security (Graham and Marvin 2001). In this way, gated "privatopias" are created where

> *"Property rights and property values are the focus of community life; and where homogeneity, exclusiveness, and exclusion are the foundations of social organization."* (McKenzie 1984).

Consequently, cities all over the world successively become divided into a mosaic of "haves" and "have-nots", not only in terms of economic and social resources but also as regards exposure to environmental degradation and subsequent threats to human health (Ravetz 2000).

4.1 Apartheid Legacy in Johannesburg

Johannesburg has been shaped by nearly one century of racially driven social engineering (see Chapter 3, Section 2.1). The racist ideology, which shaped the politics of the country until the power shift in 1994, divided the population into four main groups: African/Black, Colored, Indian and White. The apartheid politics culminated with the *Group Areas Act* of 1950, which guided spatial segregation among the four identified racial groups. Consequently, socio-functional distribution in Johannesburg, with industries and poorer residential areas located in the south and wealthier areas and commercial service in the north, was enforced and consolidated not only based on class, but on race.

In the apartheid era, the wealthy white middle and upper class thus created sparse, green suburbs for themselves along the northern and northwestern sectors, while other groups were pushed towards the southwest, on the other side of the Witwatersrand gold reef but still close to these mines (see Figure 4-4). The evacuation of Sophiatown in 1955, then a vibrant freehold for Africans and Coloreds with a prosperous music and cultural life, is a well-known example of the tough line from the authorities (Beavon 1997, Beall 2002a).

In addition to ethnic segregation, Johannesburg also exhibits stark differences in the provision and distribution of social amenities, such as hospitals, education facilities, parks, recreational areas, security provisions, and the preservation of places of cultural and historical importance. As a consequence, radical inequalities exist between different urban areas – a situation which, in turn, has induced high levels of crime, fragmented service delivery, limited socioeconomic opportunities and high poverty level in former black townships and in informal settlements (City of Johannesburg and Monitor Company 2000).

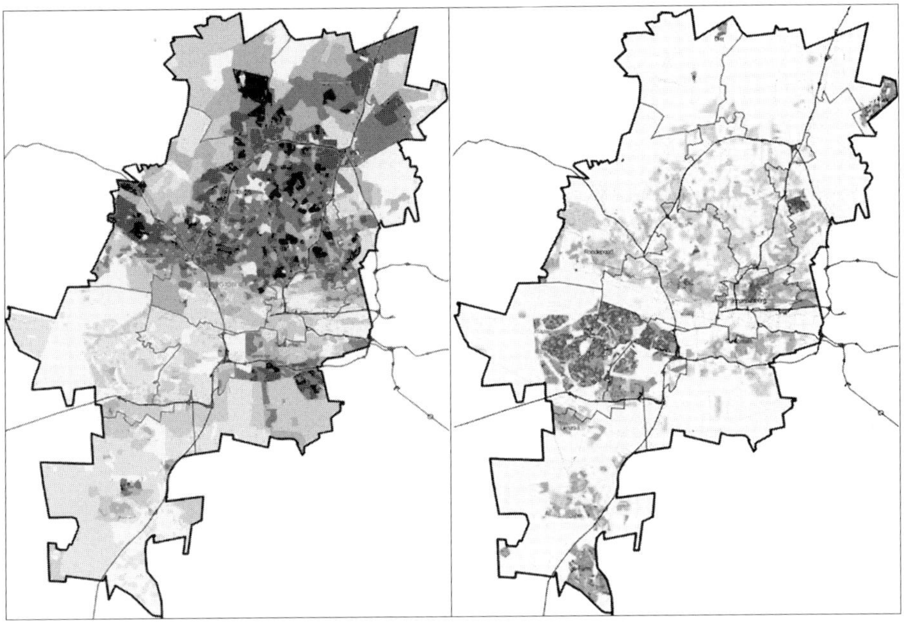

Figure 4-4. Income distribution and population density in Johannesburg. The light belt stretching east-west through the city on the population density map is the mining and industrial district located on the old gold reef. (Source: City of Johannesburg 2000)

As a result, little more than one decade after the apartheid policy was abandoned; the city is still characterized by strong racial and functional segregation. Even though initial mixing has started as more Blacks move to the inner city and the leafy, well maintained northern suburbs, most people have remained in highly segregated areas and townships (Statistics South Africa 2003).

4.2 Market-Driven Segregation in Santiago

In Santiago, there is significant socioeconomic segregation within the Metropolitan Region, where more affluent households have clustered in the eastern and northeastern segments (see Figure 4-4). In fact, the government actively eliminated illegal settlements in these areas in the late 1970's and early 1980's (Ducci 2000). This trend was further reinforced in 1981 when Greater Santiago's 17 original municipalities were subdivided into 34 socially homogeneous municipalities, delineated along rich-poor lines.

Socioeconomic segregation was further accelerated, not only by the location of continuously growing illegal settlements, but also through government efforts to provide public housing. The need to produce housing in quantity engendered construction on the cheapest land available, most often in peripheral locations, i.e. in areas where the urban poor had already settled down. Consequently, the national housing subsidy program has been one of the contributing factors to continuing urban sprawl as well as accentuating socioeconomic segregation.

Moreover, Santiago's socioeconomic segregation is also transposed onto social welfare in general, such as social well-being, education, municipal infrastructure and urban green areas, all following the same spatial distribution. The current peripheral location of low income households, thus, entrenches existing patterns of poverty and their exclusion from services – an effect, which has also been observed in Johannesburg and noted in South Africa's housing subsidy program.

5. THE SUCCESS AND FAILURE OF HOUSING POLICIES

The pursuit of adequate housing for all comprises two main elements. On the one hand, there is the matter of land tenure and, on the other, the provision of housing *per se*. In sub-Saharan Africa, for example, poor land tenure security is a prevailing issue. This constitutes a serious problem since land tenure security is one of the pillars of social sustainability (Brennan 1994).

In the case of Santiago's lower income groups, housing has always constituted a critical problem. Consequently, the poor have established themselves on peripheral lands, often within informal settlements. However, the Santiago Ministry of Housing and Urbanism (MINVU) housing programs have been successful in addressing the housing deficit, and in the 1990's, the historical trend towards a constant increase in the housing deficit was finally reversed (Ducci 2000). One reason for this accomplishment is the combination of systematic housing subsidies directed at middle income or low income families who have the debt capacity and private sector demand oriented housing production. In 2002, more than 7,000 houses were constructed in the Metropolitan Region by way of the MINVU Basic Housing Program. Even so, the situation has not been resolved with the same vigor for the poorest households. Although some governmental and non-governmental programs were initiated in the late 1990's – based on self-construction and volunteers' participation – the housing deficit remains an issue.

In Gaborone, if the estimation of the future population is correct, more than 70,000 new houses are needed in the next 20 years, requiring an annual production of more than 3,700 units (Keiner and Salmerón 2003). As Gaborone is a much smaller city than Santiago, the challenge is thus even greater. The intense demand for legal tenure and housing comprises two main challenges. First, there is the question of how to manage this amount of housing provision. Second, and more critically, is the challenge of attending to the urban poor, i.e. to secure the provision of adequate housing also for those that do not have the means to enter the existing housing market. Importantly, as discussed above in relation to inequalities in income distribution, this also comprises a gender problematic, or how to support housing ownership for single mother households. Unfortunately, as in Santiago, existing housing programs do not include the very poor. In Gaborone, the situation is also difficult since land is getting rare within the city's boundaries and servicing newly allocated plots takes time and is expensive. However, the *Botswana Housing Corporation* (BHC) and the *Self Help Housing Agency* (SHHA) are the most appropriate agencies to provide affordable housing to many citizens. Other prospects lie in encouraging public private partnerships in both land servicing and house provision.

In Johannesburg, the question of land tenure security is even more accentuated. Until 1986, black South Africans were prohibited from owning property, and leasehold and other alternatives that have existed in the past are, therefore, often associated with ethnic discrimination. As a consequence, individual ownership rights are now a matter of political redress and the government's strategies to provide housing and secure tenure are mainly based on subsidies for individual ownership (Royston and Ambert 2002).

The city of Johannesburg has mapped 89 informal settlements with an estimated 170,000 families living there (Thale 2002). There are also a considerable number of people living in illegal dwellings in backyards. Official statistics show that more than one fifth of the population lives in informal dwellings (Republic of South Africa 2003a). Accordingly, there is an urgent need for new housing projects and upgrading existing building stock. The issue is, of course, not without complications. For example, action undertaken in the dense informal settlements also involves confrontation with illegal immigrants.

The government of South Africa has implemented wide scale housing programs that, since 1994, have delivered 1.4 million units (Mahanyele 2002). In these programs, the predominant mode of delivery has been oriented to supply and individualized capital subsidies, a strategy whose effectiveness and sustainability has been questioned (Huchzermeyer 2001). However, there has been a recent local paradigm shift towards the *People's Housing Process* (PHP) (Mahanyele 2002). Whether or not this new paradigm will meet the challenges of sustainable development is still to be seen.

6. PUBLIC HEALTH

Even though life expectancy globally has increased drastically during the last fifty years, the progress has been very unevenly spread. Especially in sub-Saharan Africa, a large part of the population is severely exposed to and affected by diseases. The child mortality rate per 1,000 inhabitants (commonly used as a key health indicator) has actually increased in 14 African countries since 1990 (WHO 2003). Health issues in developing countries were more often connected to rural epidemiology, demography and administrative structures in the 1970's. A shifting focus towards urban contexts can be noticed since the mid 1980's, due to the rapid urbanization and the growing disparities in terms of environmental conditions and health care access between different population groups in the cities (Harpham and Werna 1996). The health condition is closely dependent on sanitation and access to medical services. Hence, poor households and people living in informal settlements are the most affected. One of the main outcomes of modern city planning as it took form in Europe during the 19th century, was that it provided clean drinking water, sewage and waste management and thus drastically improved the health conditions for the poor households in the rapidly growing industrial cities of the time. Diseases such as cholera and tuberculosis, which were virtually eradicated in cities of Europe and North America after industrialization, are still common causes of death in many cities in developing countries. In the case of Gaborone, the use of pit latrines and overcrowding in some areas (e.g. Old Naledi) cause health problems, and pollute groundwater. Overflowing sewage ponds endanger both the ground and superficial water, like the Gaborone dam or Notwane River. There is an urgent need to connect all plots to adequate water and sewage services. Wastewater can, if recycled in a correct way, be reused for ground water recharge, potable water or irrigation. Since the water situation in Botswana is not the best, research and training in water recycling must have a high priority. The storm water drainage system of Gaborone is inadequate to handle high quantities of seasonal storm water. The open channels are often filled and blocked with mud, sand or rubbish, which leads to recurrent street floods. The city authorities cannot cope with these problems because of the lack of funds and skilled manpower. Thus, health conditions have a clear connection to economic situation, both individually and on a national scale, but this relation is not solely dependant on finance.

The *World Health Organization* (WHO) emphasizes the crucial role of a strong health administration system based on primary health care and integrating acute treatment and preventive measures. There is also the question of non-transferable diseases, like tobacco related cardiovascular disease, a large set of diseases related to air pollution and other environmental pressures and mental illness related to stress or social pressures, which are often connected to industrialization and economic growth (WHO 2003).

One of the most overwhelming threats to sustainable development, especially in Southern Africa, is the pandemic of HIV/AIDS. In South Africa, a fifth of the population is infected but in Gaborone the situation is even worse: 38.8% of the population was estimated to be living with HIV/AIDS in 2001 – the highest prevalence worldwide (CIA 2003). Current UNAIDS figures show prevalence rates among those aged between 15 and 49 of 44% and more. In Gaborone, HIV/AIDS is recognized as one of the most severe problems. There has been a drastic fall in the life expectancy, expected to drop down to 42 years by 2010 (Government of Botswana 2002a).

However, even if the situation is being discussed daily in the media, and anti-HIV/AIDS programs are implemented countrywide, people do not appear fully conscious of the future impact of AIDS on Botswana's society, i.e. on its productivity and its labor force. The problem is difficult to address, not least because the HIV infection can go unnoticed for a long time before AIDS manifests itself. Poverty is an important contributing factor, both to the spread and the effects of the disease. Many women are forced into prostitution and poverty and lack access to information and prevention methods. Consequently, the prevalence of the disease is unequally distributed among different groups, as regards race and ethnicity, age and gender – the worst affected group in South Africa being black women 25–29 years old (City of Johannesburg 2002b).

In 2003, the South African cabinet approved the *Operational Plan for Comprehensive Treatment and Care for HIV and AIDS* (Republic of South Africa 2003b), which among other things include the nationwide distribution of antiretroviral drugs. This decision was warmly welcomed by many organizations working with HIV/AIDS issues and the new government policy gives more hope to the millions of South Africans affected by the disease. There is now reason to hope for change in the near future, as the problem has been well recognized at all levels of the South African society. However, as women and children are the most exposed groups, a main challenge still lies in empowering women to change their current socioeconomic circumstances.

In the Johannesburg Metropolitan Area, the situation is somewhat better than nationwide statistics indicate, since it is estimated that just about 10% of the population have HIV/AIDS. However, when taken into account that 30% of pregnant women are infected, it is obvious that AIDS related diseases will prevail as the greatest cause of mortality far in the future (City of Johannesburg 2002b). This is an issue that not only affects the direct victims and their families but also causes huge damage to the economy and the society as a whole due to loss of labor and skills, changes in household structure and destabilization of families, increased infant mortality rates and pressure on health facilities and government resources (City of Johannesburg 2002a).

7. INFRASTRUCTURE AND TRANSPORT

In low and middle income country cities, the mismatch between high rates of urbanization and inadequate governance capacity is reflected in the poor state of their infrastructure and services (El-Shakhs 1997, Rakodi 1997b). The City of Gaborone, for example, is plagued by the lack of managerial capacity especially for high caliber administrators and technical staff in fields like engineering, architecture, project management etc. Cities can function socially and economically only if they have adequate and efficient infrastructural services in key areas, such as drinking water, sewage, roads, transport, electricity, telecommunications, and waste management. The development and efficient operation of urban infrastructure, thus, constitute an essential ingredient for a city's livability, attractiveness and competitiveness.

The variations in levels of infrastructure and services fall into three major categories (Wekwete 1997):

- Urban development without any basic infrastructure services, i.e. in informal or squatter settlements
- Formal development projects, but with inadequate infrastructure networks
- Formal urban development supported by basic infrastructure networks but sustaining poor levels of service

The first category constitutes the most serious threat as it challenges formal requirements of legal tenure before linking up with municipal infrastructure. Moreover, the inability of affected households to meet costs for network development and operation poses the additional challenge of how to finance necessary infrastructure. The formalization of tenure linked to the upgrading of infrastructure services is therefore critical for sustainable urban development in such areas.

The second category is characterized by development on legally owned land but without developing adequate infrastructure networks to match the increasing intensity of land use. Although developers and owners may be in a position to pay for infrastructure services, incapacity for planning and implementation at municipal levels hinders the required development and management.

The final category is common in low and middle income country cities where areas with sufficient infrastructure networks consistently experience poor service levels, often characterized by interruptions, overloading, long periods of breakdown before repairs etc. Ultimately, breakdowns in services contribute to environmental impacts, such as water loss due to network leakage, flows of raw sewage into storm drains and rivers and illegal solid waste dumps.

Transport systems are basic infrastructure networks with immense importance for numerous urban functions as well as for the urban environmental quality. In the *Greater Gaborone Region*, the existing mass transportation system allows people to live in the surrounding peri-urban areas and to work in the centre of the city. It is completely run on a private basis with mini buses (10 to 20 passengers). These buses – often in bad technical condition – ensure commuter traffic needs on main roads, but there is no overall coverage for transportation. Consequently, the growth in population, in combination with an increase in private car ownership, results in constantly increasing traffic volumes on the roads. This has resulted in severe traffic congestion during peak hours in some places of the city's network.

A publicly procured mass transport system would constitute the core of an efficient, environmentally friendly and safe Gaborone transportation network. However, the Gaborone terrain is completely flat and, thus, very suitable for bicycle use. Even so, about 30% of individual transportation is by foot and almost none by bicycle (Government of Botswana 2002b). One reason for the absence of bicycles in the Gaborone streetscape may be that the road system is not very safe due to reckless driving and a lack of maintenance. A challenge is hence to facilitate safe walking and cycling as a complement to mass transport means, for example, by planning for suitable pedestrian and bicycle paths.

In Johannesburg, the City operates a municipal bus system and a separate private bus company serves the connections between Alexandra, the City Centre and Soweto as part of a national transport subsidy program. These buses amount to 13% of passenger transport but as much as 29% is provided by the burgeoning mini-bus-taxi industry (City of Johannesburg and Monitor Company 2000). A rail system connects to some townships and surrounding cities but there is no subway system or trams in Johannesburg. Although the mass transport system is better coordinated in Johannesburg than in Gaborone, it still fails to provide adequate public transport means – a shortcoming that is aggravated by pressure from the uncontrolled growth of commercial centers and high-density informal settlements as well as poor relationships between housing delivery and job opportunities (City of Johannesburg and Monitor Company 2000). The inhabitants therefore view the private car as the ideal transport mode if one has the economic means. As a consequence of the resulting traffic congestion, the average daily commuting time is estimated at 72 minutes, in spite of the relatively modern road network (City of Johannesburg 2002b). Moreover, the overall low accessibility for pedestrians forces many people to walk on traffic lanes, to jump fences and to cross thoroughfares at inappropriate places.

In Santiago, rapid economic growth, i.e. growing population, higher incomes and more trips taken, have caused an explosive increase in private car use. Currently, the motorization rate is of 135 vehicles per 1,000 inhabitants, causing the urban transport system to deteriorate. Moreover in 2010, the current 29% automobile share in the modal split of metropolitan transportation is estimated to grow to 60% (Malbrán 1998). However, a number of initiatives have recently been taken to improve the situation, such as the extension of the Metro subway, a sizeable program for urban and regional highway modernization and a comprehensive plan for coordinating the metropolitan transportation system.

8. ENVIRONMENTAL IMPACTS

8.1 Ecological Footprints

In addition to severe socioeconomic repercussions, poverty and low standards of living for many households in low and middle income countries have two major environmental implications.

First, due to their limited consumption, their average ecological footprint (see, for example, Redefining Progress 2003) is extremely small compared to those of high income country households. This means that, on a global environmental scale, poor populations pose comparatively low risks and concurrently offer minimal opportunities for reducing impacts in areas such as climate change or the management of production/consumption issues. However, as their income levels and standards of living improve – one of the key goals of sustainable development – energy consumption, industrial products and market-based services will increase and ecological footprints will grow bigger. Consequently, there is a significant challenge of how to keep the ecological footprint small as income levels and the standard of living improve (Sustainable Cities 1992).

Second, even if resource consumption is low, the lack of adequate infrastructure and economic means for environmental management will still result in stronger pressures on the local environment in cities of low and middle income countries than in high income countries. If nothing is done, air, water and solid waste pollution as well as extravagant land use and urban sprawl will undermine their livability, attractiveness and competitiveness. Unfortunately, even though such local impact issues are often understood as a relevant agenda for local governments, the lack of institutional capacity often limits the extent to which municipalities succeed in managing the impacts (Sustainable Cities 2000).

Within this context of potentially growing ecological footprints, on the one hand, and threats of increasing local environmental degradation on the other, the common pattern in low and middle income countries has been to focus on the reduction of local environmental impacts while leaving aside the issue of resource consumption and environmental impacts on the global scale. The global environmental agenda is rather seen as the domain of the rich, with their resources for intricate environmental management and access to frontline clean technologies. For low and middle income countries, global environmental issues might only attract effective attention after basic needs have been significantly addressed. Put differently, the global challenge is often seen as too remote to justify a release of public and private sector resources, given current more pressing needs in relation to poverty.

Consequently, issues like the conservation of bio-diversity, renewable energy, water conservation and waste reuse/recycling do not receive significant attention in low and middle income countries – regardless of whether they are discussed at state or city level. This is not a trivial issue as it accounts for key differences in the interpretation of the sustainable development paradigm between stakeholders in low, middle and high income countries, respectively.

8.2 Air Pollution

Santiago's most burning environmental issue is air pollution. The city suffers under the worst air pollution levels in Latin America, together with Mexico City and São Paulo. Much of the pollution stems from the city's topography and climate, rapid urbanization, industrialization and motorization and has serious impacts on human health. Even though greenhouse gases, global warming and climate change may still be a remote environmental issue for residents and authorities in Santiago, both parties understand the seriousness of local air pollution (see Figure 4-5).

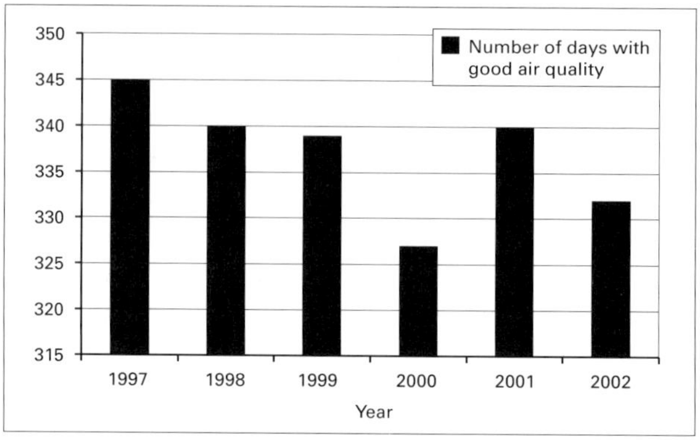

Figure 4-5. Air pollution in Santiago de Chile 1997–2002 (Source: Demarmels 2003)

Since buses and private cars are by far the most common modes of transport in Santiago, the transportation system is the principal source of air pollutants (PNUMA-IEUT 2003). Although the pollution levels can be strongly linked to urbanization rates and transport, the problem is exacerbated by geographic conditions. Due to its location in an isolated aerial valley with limited natural ventilation and high atmospheric stability, air pollution comes in different forms and their impact is heightened. In addition, thermal inversion triggers the emission, dispersion and deposition of different air pollutants during the fall and winter. Atmospheric contamination is composed of 80% of suspended dust resulting from deforested slopes, unpaved streets, and alluvial mud. Other more harmful contaminants in the cloud of smog include micro particles resulting from diesel oil combustion produced by the numerous buses that circulate daily through city streets. Future development will have to take all of these aspects of the air pollution problematic into consideration. In Chapter 5, Section 2.1, a strategy for minimizing air pollution impacts in Santiago is shown.

In comparison, Gaborone is not much affected by air pollution. Even if some roads are congested during rush hours, overall traffic volumes are still moderate. However, as car ownership is increasing rapidly, growing numbers of commuters and the rapid increase in dispersed shopping malls will induce deterioration in air quality. Another source of air pollutants are the numerous burrow pits in the back yards where rubbish is burned. In addition, many citizens – mostly the urban poor – use wood for cooking, which not only pollutes the air but also threatens respiratory health and the vitality of the forest's ecosystems.

8.3 Urban Environment, Open Spaces, and Waste in Gaborone

As aforementioned, Gaborone is facing quite a number of environmental problems that include poor waste disposal and littering, pollution and traffic congestion, flooding, soil erosion etc. In response, the city has embarked on environmental management programs supported by the central government. Through these programs, the city has factored environmental issues in its planning programs and audits its activities to ensure sustainability. Major achievements of the audits have included the identification of environmental problems facing the city and the development of sector environmental profiles.

Gaborone was once planned as a Garden City (see Chapter 3, Section 3.3). A city with such an image may be attractive, however, it might not be the right concept in a country regularly afflicted by drought. In addition, many of the city's open spaces are frequently abused as dumping sites, driving schools, urban agriculture or other uses. But these open spaces are essential for maintaining biological and hydrological balance and could create better climatic conditions and give the city a more friendly appearance. As the city grows, such green areas are becoming more and more important, and their protection, maintenance and upgrading are a matter of urgency. The Gaborone dam and its surrounding green buffer is one of the largest and most fragile ecosystems in the Gaborone vicinity, which must be protected from uncontrolled physical development and water pollution.

In Gaborone, there is a huge number of illegal dumping sites and burrow pits within the city's perimeter, especially in the urban-rural fringes. This is both a serious environmental threat and a visual intrusion to the city's image. Littering devaluates areas of natural beauty with their potential for passive recreation. The separation of different types of solid waste is not common in Gaborone. Although about one third of Botswana's aluminum cans are recycled, there is still a huge lack of education, awareness and governmental initiative concerning this subject. What is happening at the moment is that all sorts of waste are collected and brought to the landfill. By recycling paper, glass, iron etc, Gaborone could create new jobs and simultaneously protect the environment (see Chapter 7, Section 1.1).

Table 4-1 provides a list of environmental problems facing Gaborone City. The method used, particularly the presentation of the problems on sectoral basis, can be implemented by physical planners to depict the situation in their planning areas. It is important to note that the environmental problems identified are area specific. Similarly, the solutions should be local and relevant (Government of Botswana Environmental Planning Programme 2001).

Table 4-1. Selected Environmental Problems in Gaborone, 2001

Sector	Environmental problems, issues and concerns	Potential solutions
Transport and communication	Congestion Poor pedestrian paths	Staggered working hours Include pedestrian walkways in all road designs
Building and landscape issues	Excessive land clearance Construction waste Unmanaged open spaces	Strict enforcement of regulations governing dumping Encourage communal management of open spaces
Industrial and commercial development	Air and water pollution Generation of industrial waste exposure to industrial hazardous waste Use of leaded fuel	Promote measures aimed at proper waste management Support NGOs that promote these measures
Forestry and horticulture	Overexploitation of wood resources	Encourage tree planting and development of woodlots
Population dynamics	Overcrowding in some parts of the city Use of pit latrines Congestion of informal sector traders Squatting	Adoption of innovative housing strategies Partnerships with private sector to increase supply of land; work with informal sector traders to address congestion
Energy resources	Use of firewood	Encourage the use of alternative energy sources, for example coal
Urban Agriculture	Use of pesticides Use of chemical fertilizers Poor siting of agricultural and horticulture activities	Promote the use of organic fertilizers Provide guidelines for the location of agricultural and horticulture activities
Public health	Indiscriminate waste disposal Poor waste collection by local authorities	Promote proper waste disposal Encourage community-based low cost waste collection strategies
Mining and quarrying	No proper sites for extraction of sand and gravel Poor rehabilitation of land Destruction of natural landscape	Work with local authorities to designate official sites for gravel and sand extraction

Source: Adapted from Government of Botswana EPP 2001

9. GOVERNANCE

9.1 The Challenge of Centralization

The challenges analyzed in previous sections clearly demonstrate the critical leverage of sound and efficient urban governance for turning these challenges into opportunities for sustainable urban development. But what characterizes such an urban governance system?

First, although many development issues may be efficiently addressed on more central levels, there are also many questions that need local adaptation. However, institutional structures in developing countries are often strongly linked to the governance of the state, such that local authorities are seen as third or fourth level in the government hierarchy (Sustainable Cities 2000). This relationship significantly affects the activities of local authorities, both in terms of policy formulation and resource allocation. The options for local municipalities to elaborate particular development agenda turn out to be limited, because in the end, they act as mere translators of centrally approved policies.

More decentralized approaches to urban governance would confer better possibilities in terms of generating local resources and for envisioning short and long term outcomes of local development initiatives (see Chapter 7, Section 2.3). However, both small and large municipalities would also face a capacity challenge in terms of responsive management systems as well as adequate human resources.

Second, structures with different scales and extents exist in the human society as in ecosystems, but these socio-technical configurations rarely follow the same boundaries as formal administrative structures (Kain 2003). A core challenge for urban governance is, therefore, to identify relevant groups of stakeholders and bring them together in collaborative planning and decision-making processes.

Third, as claimed in *Agenda 21*,

> *"One of the fundamental prerequisites for the achievement of sustainable development is broad public participation in decision-making."* (UN 1992: Chapter 23.2)

Such enhancement of the direct accountability of municipal authorities to residents and stakeholder groups entails an important challenge for urban governance systems to facilitate efficient and democratic ways to involve individuals as well as organized groups of people. This constitutes a particular challenge when it comes to groups with traditionally weak influence, such as women, the young, urban poor, immigrants and ethnically discriminated groups. In addition to obvious ethical imperatives, these groups may also be carriers of local knowledge with high relevance for the planning of sustainable urban development.

Ultimately, the call for decentralization, stakeholder involvement and broad accountability challenges municipal authorities to develop transparent and efficient communication systems. The function of these systems is to support information sharing and feedback, as well as procedures for representing different stakeholders in relation to various decision-making processes. However, since this approach is very different from the prevailing technocratic program planning and implementation procedures, it will take time for principles of good practice to emerge (see also Chapter 5, Section 5.2).

9.2 City-Regions and Municipal Structures

The three case study cities display quite different conditions in their institutional setup. Chile as well as South Africa uses a three-tiered government system, but Chile's local municipalities are much smaller, and as a result, Santiago consists of 34 autonomous local bodies. Therefore, urban development issues for the whole city must, therefore, be handled by the regional or national governments. This is not without complications, as the local authorities possess a strong position as independent entities. In Santiago, there is also a wide range of governmental bodies on central and regional levels with responsibilities in the urban planning field. Unfortunately, there exists a severe lack of coordination between them (Programa Chile Sustentable 1999). Currently, there exists no joint vision for sustainable development in Santiago. Instead each ministry formulates its own vision and programs independently. The need for improved coordination among authorities is further highlighted in Chapter 7, Section 2.3.

Nonetheless, there are some recent attempts to develop broader planning instruments, for example the *Regional Master Plan* and the multi-corporate plan for the transportation system. However, conflicts continue to emerge from the overlapping of municipal bodies and the absence of an effective and efficient Metropolitan Government. New plans must pass through a complex legal system before implementation and rapid changes cannot take place. Even so, although Chile's tradition of strong legislation reduces options and thus hinders sustainable urban development in some regards, it also enhances possibilities of implementation when the relevant actors finally validate plans.

Like Santiago, the metropolitan area of Johannesburg used to be divided into several autonomous municipalities. The institutional system inherited from the apartheid regime was, thus, characterized by complexity, fragmentation and overlapping functions (see Chapter 7, Section 2.3). In 1994, the new democratic government had to start more or less from scratch with the formulation of a new constitution, new legislation in all areas and the development of new administrative systems. After these institutional reforms, Johannesburg is now acting as a "unicity" with only semi-independent city regions. Even so, the planning system still suffers from a lack of integration of strategies between the different departments and areas of legislation.

Also in terms of public participation (see Chapter 7, Section 1.7), the ambitions have been high to create a new democratic society without former discrimination and inequalities. Consequently, the new institutional system in Johannesburg provides a set of opportunities for citizen participation and consultation in planning and decision-making processes. The City Council of Johannesburg has decentralized its administration into eleven local regions and 109 wards, each with its own elected ward-councilor. Through the ward councilor, local citizens have a direct representation in the City Council. Very important for the grassroots democracy is also the overall community consultation for the *Integrated Development Plan* (IDP; see Chapter 5, Section 6.2) carried out at ward and regional levels. There have also been several processes to develop a joint vision, which have equally involved a broad set of diverse stakeholder consultations. The resulting document, *Joburg 2030* (City of Johannesburg 2002b), now works as a foundation for planning and decision-making in the city (see Chapter 5, Sections 1.5 and 7.1).

As Botswana is a much smaller country, the governmental system is comparatively less complex. Gaborone City constitutes a single local municipality. However, the urban region is larger and the peri-urban villages or satellite settlements, which definitely form a critical component of Gaborone's urban system, thus fall outside the municipal boundaries of Gaborone. As a result, a *Regional Physical Development Plan*, involving also the surrounding municipalities should be produced. In 1994, a *Greater Gaborone Structure Plan*, providing a framework for integrating economic, social, institutional and physical development within an areas of about 1,000 km^2, had been worked out, but never approved.

Although basic institutional structures are in place in the planning systems of all three cities, the synchronization and monitoring of the various plans is still missing. However, like Johannesburg, Botswana (and by implication Gaborone) has gone through a broad visioning process presented in the document *Vision 2016* (Presidential Task Group 1997). This national vision is a cross-sector outline of the country's future aspirations, encompassing social, environmental, political, cultural and spiritual aspects.

10. FINANCING

10.1 Weak Revenues and Weak Financial Management

Many initiatives required for sustainable urban development are likely to go unimplemented due to insufficient funding, especially in low and middle income countries. This problem may manifest itself in different ways, such as a weak and unevenly distributed revenue base, inadequate grants from national or regional governments, a weak budget balance between capital and operational expenditures, inappropriate financial management systems – coupled with low credit ratings.

The weak revenue base in many developing countries is due to several factors. First, it is a consequence of the inability of most residents to pay rates and service costs. Municipalities have to rely on a very narrow range of properties actually paying these fees (Rakodi 1997b). Even if there exist wealthy areas with a stronger revenue base and other resources for development, it may be difficult for urban planning authorities to direct these recourses to areas where it is most needed.

Second, in cities under strong central government control, most financial decision-making takes place in the ministries of the central government. As a result, such cities are not fully responsible for generating their own revenues, since most of their budgetary requirements are met through grants and project funding from the central government. Unfortunately, such grants are often very limited compared to existing demands in relation to infrastructure development and services, especially where a significant proportion of the population is living in poverty (Rakodi 1997b; see also Figure 4-6).

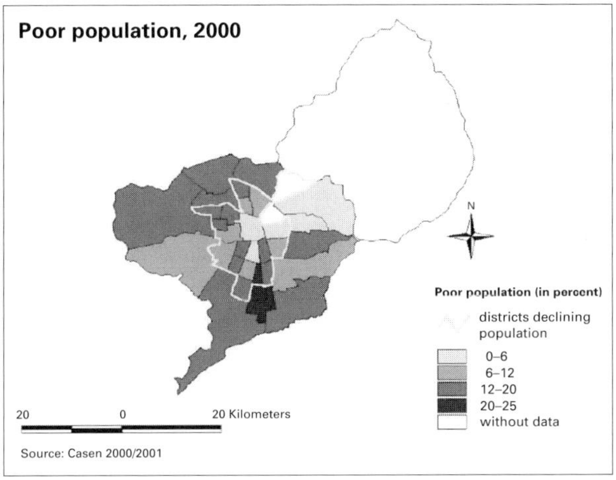

Figure 4-6. Poor population in Santiago de Chile (Source: Gobierno de Chile 2002)

Third, budgets of most developing country cities are dominated by operating costs, with the remuneration of municipal employees as the main expenditure. Therefore, capital spending on new infrastructure or renovation of existing networks often takes place on a crisis management basis rather than as a conscious strategy. Over time, the result is a continuous downward trend in network efficiency and in the urban system performance as a whole. This, in turn, feeds into undermining a city's competitiveness and attractiveness in relation to investment and economic growth (Wekwete 1997).

Fourth, local financial management in developing countries is often characterized by weak accounting and reporting systems. Moreover, in relation to the sophisticated nature of financial operations in large municipalities, most of them are unable to attract and retain adequately qualified personnel. As a consequence, local financial management is often open to abuse through avoidable errors, mismanagement, fraud and corruption (Wekwete 1997).

A weak revenue base and weak accounting systems severely undermine credit rating with major credit institutions and especially development banks. In addition, most national governments place restrictions on municipal government borrowing. Finally, when it comes to more innovative financing mechanisms such as public-private partnerships or environmental impact levies, the complexity of such measures – coupled with a lack in skills and expertise in municipal governments – has limited their implementation in developing country cities (Wekwete 1997).

10.2 Municipal Autonomy and Redistribution

Gaborone serves as a good example of a city where the central government plays the major role in planning and financial decisions (see Chapter 7, Section 2.3). Like most municipal authorities, the Council gets the bulk of its resources from the central government; on average the city gets a 100% grant for its development budget and over 60% of its recurrent budget from the central government (Mosha 2002, Government of Botswana 2002c). Income from its own resources is very limited indeed and the annual budget clearly does not meet the city's needs. The city hardly sources money from the open market in the form of loans, as it is not allowed to do so. The City Council should be encouraged to generate more funding through its own resources to support essential development projects.

The Council has to perform certain tasks regarding service provision. These include housing and land servicing; infrastructure, such as sewerage schemes, primary and secondary roads, power and water supply will be required to expand, augment or rehabilitate existing services. However, the Council most certainly does not have the financial, human or physical resources to undertake the maintenance and coordination of the completed services listed above, and certainly not the capacity to implement the capital expenditure projects.

Johannesburg, in contrast, generates up to 95% of its budgetary requirements internally. The city also enjoys great autonomy with regard to external borrowing as well as the implementation of innovative financing strategies, such as public-private partnerships. Grants from central and provincial governments are often targeted at specific delegated programs or projects (City of Johannesburg 2003).

In terms of budget financing, Santiago is similar to Gaborone. The Chilean national government accounts for 95% of all taxes collected nationally and also redistributes a large share of these funds to regional and municipal governments (Barra and Jorratt 1998). Moreover, the national government sets the rules by which municipal governments can raise revenue. It may be argued that the regional level would be the natural body for urban planning in Santiago, as it covers the actual city region. However, the regional government has virtually no independent revenue-raising capability and, thus, weak possibilities to act as a coordinating financing body.

When it comes to the question of redistributing resources *within* the city, Gaborone does not seem to constitute much of a problem due to its reasonable size. Santiago, however, is quite another issue. Here, the problem of redeploying resources within the city is obvious, since the local municipalities – demarcated in accordance to patterns of different income levels – have a high level of financial independence. Consequently, municipalities in poorer districts must rely on additional funding through the Municipal Common Fund allocated by the national Ministry of Finance.

In Johannesburg, the post-apartheid institutional system was designed to bridge the gap between poorer and wealthier communities within the city. In a first attempt, four councils were set up, integrating former white and black communities with different economic status. This system was rearranged, however, due to critical financial problems, and the "unicity" was created, where the whole city shares a common tax base (Beavon 1997). This provides for the City Council to direct development resources to areas where the need is most urgent although the available funding for such purposes is still limited.

11. CONSOLIDATION AND CONCLUSIONS

In Botswana and South Africa, having no job is a brutal reality as a large part of the population (close to about 50%) lives below the poverty line. Although the overall levels of poverty in Santiago have been reduced with the ongoing economic growth, the income inequality is still alarmingly high. The contribution of unemployment to poverty as well as to disparity in the access to municipal services is often pointed out as a major challenge for low and middle income classes in cities of developing countries. This challenge also includes secondary manifestations, such as high illiteracy levels, poor environmental health, inadequate shelter and services.

There is also a gender aspect to poverty, since women tend to take greater socioeconomic responsibility of the household, and yet they do not enjoy similar access to opportunities as men in the labor market. The challenges in relation to poverty are threefold: opportunities for employment and self-employment must be ameliorated; income disparities must be addressed; and the status of women in society must be improved, so that they may contribute to personal and national economic development.

Unfortunately, it appears difficult to find feasible developmental or wealth distribution mechanisms to reverse the stark disparities in poverty and income inequality within acceptable timeframes (Sustainable Cities 2002). Although short-term cross-subsidization for access to services and amenities is commonly practiced, this requires implementation at a large scale and over several generations in order to overcome the structural constraints that reinforce the vicious poverty cycle. Therefore, it is likely that endemic poverty and disparities in distribution of income will remain one of the most persistent challenges for sustainable urban development. Acceptable strategies for the increased creation of wealth complemented with more equitable distribution mechanisms are urgently required for this challenge to be effectively addressed. Economic growth alone has proved to be inadequate (Sustainable Cities 2002).

In all three countries, the case study cities enjoy a primary-city status, e.g. Santiago is home to as much as 40% of Chile's population. Such primacy plays a significant economic role, not only in terms of the sheer size of the market for basic products and services, but also in terms of providing the largest market for luxury households as well as commercial goods and services. Even Gaborone and Johannesburg, with a smaller share of the national population (11% and 8% respectively), have critical roles in the economic development of their respective countries. The primacy status thus escalates the city's attractiveness for people searching for jobs and other opportunities for livelihoods.

The case study cities are actively dealing with the problems of urban sprawl, even though the initiatives have still not been very successful. Urban sprawl occurs at different rates related to the population growth and stage of development. The driving forces are rural-urban migration and dynamics related to land price, access and systems of tenure. Constraints in the management of urban sprawl also arise from institutional settings as well as inherited cultural values and practices of the people. In cities like Gaborone, originally planned after garden city ideals, or Johannesburg, with large townships based on principles of "match-box" single unit mass housing and large suburbs of garden villas, plot intensification and densification appear to be essential.

However, it is also necessary to relate to the ideals of the *green city*. Urban open spaces are essential for biological and hydrological balance; they create better climatic conditions and give the city a friendly appearance. Furthermore, it is crucial to consciously plan and design the rural/urban interrelation, i.e. the shape of the urban boundary. This challenge encompasses a number of issues, such as how to maintain and utilize peri-urban open land, how to maintain the identity and capacity of surrounding towns, and, subsequently, how to locate areas of new housing so that they support the idea of urban boundaries and urban identity.

Socioeconomic and functional segregation is apparent in the case study cities just as in many other cities of the world. The reasons are often a combination of market forces and conscious planning regulations. It may be argued that weak political will and inadequate governance capacity (institutional structure and policies as well as human and financial resources) has hindered innovativeness in terms of responsive policies, legislation and programs for effectively addressing such segregation. Zoning is still a strong modern planning tool, although it has been questioned in the contemporary planning discourse as it inhibits mixed-use and livable urban landscapes. In addition to the inequalities and social conflicts involved, socioeconomic segregation prevents the functional integration of social and technical infrastructure especially for poor areas where people cannot contribute to the investment costs.

Socioeconomic and functional segregation in the case study cities is rooted in different historical and contemporary conditions. Central Gaborone was originally built as the capital city within a limited geographic area containing mostly government facilities and related housing and services. Recent rural-urban migration has resulted in a rapidly growing need for housing, which has proved impossible to manage within the planned area. Instead, new settlements have grown on private and tribal land where legislation and control is out of the hands of the city administration.

In Johannesburg, the socio-spatial manifestation of the apartheid legacy still poses a difficult burden in planning and city management. Continuing ethnic and socioeconomic segregation – coupled with the decaying inner city – does not at all look promising for the sustainable development of the city.

The public housing policy of Chile has been declared a success. However, when the housing policy is examined against the issue of socio-spatial segregation in Santiago, it clearly demonstrates the contextual challenge involved. The housing subsidy programs in Chile and South Africa have tended to reinforce the spatial segregation of low-income households. This is due to the tendency to locate such housing on the cheapest land in an attempt towards spreading the benefits to the largest number of citizens.

In an open market system, such land will generally be found in areas unsuitable for residential development or in the urban periphery, far away from established and highly attractive nodes. Consequently, the sought outcome of conferring economic benefit to the poor through shelter and secure tenure is often not achieved. Access to other amenities and opportunities, such as schools, health clinics, infrastructure, recreation, jobs and markets, is undermined by such locations.

Another related challenge concerns emissions and air quality problems of the cities. In Santiago, fixed source-related emissions dropped significantly between 1997 and 2000. This was due to cleaner technologies, such as the installation of filters and emission cleaning plants, especially following the enforcement of environmental legislation on emissions control by health authorities. However, this contrasts with mobile source-related emissions mainly from transport, which increased significantly over the same period. Although Gaborone seems to enjoy low pollution levels, it requires the formulation and enforcement of responsive policies and legislation to mitigate escalations as urban development continues.

The existing governance and institutional structures of the cities have very different historical and cultural legacies. Efforts to manage inherited, contemporary and future challenges result in continuous initiatives to change and redesign legislation and administrative organization. Crucial to sustainable urban development is the coordination between levels of government and cross sectors in public administration as well as the cooperation with the private sector and local communities (see Chapter 5, Section 5.4).

To efficiently address the local problems, there is a need for strong local governments and efficient local economic systems for the fair distribution of resources in line with the vision of sustainable urban development. However, it is critical that decentralization initiatives are complemented with capacity building programs. Although increasing levels of municipal financing autonomy may be required for sustainable urban development – as it enhances the accountability of municipalities to their various stakeholder groups – the viability thresholds in terms of urban population and economic size are difficult to anticipate.

The case study cities show general conformity towards trend escalation in relation to critical issues and challenges resulting from unsustainable urban development practices. Key among these include urban sprawl and related implications on property ownership, low density, inadequate thresholds for public transport and amenities, poor integration and socio-spatial segregation.

With the exception of Santiago where unemployment and poverty levels have stabilized, the other two cities show trend escalation in socioeconomic aspects. For example, the unemployment level and the role of informal employment are extremely high and have been escalating in Gaborone and Johannesburg. Consequently, inequality in income distribution (especially across gender lines) is still high in all three cities and has been escalating in Gaborone and Johannesburg. Sustainable urban development for Gaborone and Johannesburg is further threatened by the escalation in the impacts of the HIV/AIDS pandemic, especially among the economically active population groups.

The state of governance in the three cities has started to respond to these issues with policies and programs, legislation and institutional structures, which are developed interactively across the various spheres of government. However, all three cities show varying levels of inadequate capacity and political will to align and coordinate urban development initiatives with such legislation and policies. In Santiago, for example, weak co-ordination between the various local municipalities, regional and national governments tends to undermine effective implementation of land use and environmental policies as well as enforcement of related legislation. In Santiago and Johannesburg, the housing subsidy programs have been implemented in a way that undermines urban integration and socioeconomic objectives of socio-spatial integration across income levels even though they contribute towards stabilizing housing backlog and reducing poverty.

In all three case study cities, the socioeconomic dimension of sustainable urban development dominates over the environmental issues. Instances where environmental issues get prioritized often relate to city-specific impacts such as smog in Santiago and loss of arable land in the villages around Gaborone. Global environmental issues such as global warming, water scarcity, managing overconsumption and overproduction by the wealthy is rarely emphasized through the implementation of policies, programs or legislation by these cities or their respective countries. Consequently, the expectation that developing countries will catapult themselves into cleaner technologies (smaller ecological footprint with increasing standards of living) through sustainable development cannot be justified on the basis of the prevailing trends in the case study cities at this point. It is therefore clear that efforts to address sustainable urban development in a comprehensive manner will require many more resources and several times the capacity currently in place within such cities.

REFERENCES

Barra, P. and M. Jorratt 1998. *Estimación de la Evasión Tributaria en Chile.* Estudios Económicos. http://www.sii.cl/aprenda_sobre_impuestos/estudios/tributarios13.htm, retrieved on March 29, 2004.

Beall, J. (Ed.) 2002a. *Uniting a Divided City: Governance and Social Exclusion in Johannesburg.* London: Earthscan Publications.

Beall J. 2002b. Globalization and Social Exclusion in Cities: Framing the Debate with Lessons from Africa and Asia. *Environment and Urbanization.* Vol. 14, No. 1.

Beavon, K. 1997. Johannesburg: A City and Metropolitan Area in Transformation. In C. Rakodi (Ed.) *The Urban Challenge in Africa – Growth and Management of its Large Cities.* Tokyo, New York, Paris: UNU Press: 150–191.

Behrens, R. and P. Wilkinson 2003. Housing and Urban Passenger Transport Policy and Planning in South African Cities: A Problematic Relationship? In P. Harrison, M. Huchzermeyer and M. Mayekiso (Eds.) *Confronting Fragmentation: Housing and Urban Development in a Democratising Society.* Cape Town: UCT Press: 154–174.

Brennan, E. 1994. Mega City Management and Innovation Strategies – Regional Views. In R.J. Fuchs, E. Brennan et al. (Eds.) *Mega City Growth and the Future.* Tokyo, New York, Paris: UNU Press: 233–255.

Catholic University of Chile 2004. *DIMSUD Synthesis Report: Towards Sustainable Urban Development in Santiago, Chile.* Unpublished working paper.

CDE Centre for Development and Enterprise 2002. *Johannesburg: Africa's World City – A Challenge to Action.* Johannesburg.

CIA 2002. *The World Fact Book 2002 – South Africa.* Central Intelligence Agency, United States Government. http://www.cia.gov/cia/publications/factbook/geos/sf.html, retrieved on April 4, 2003.

CIA 2003. *The World Fact Book 2003 – Botswana.* Central Intelligence Agency, United States Government. http://www.cia.gov/cia/publications/factbook/geos/bc.html, retrieved on March 12, 2003.

City of Johannesburg 2000. *State of the Environment.* http://ceroi.net/reports/johannesburg/csoe/Default.htm, retrieved on March 3, 2004.

City of Johannesburg 2002a. *Integrated Development Plan, 2002/03*. Johannesburg.

City of Johannesburg 2002b. *Joburg 2030 – Full Report.* http://www.joburg.org.za/joburg_2030/2030_documents.stm, retrieved on March 2, 2004.

City of Johannesburg 2003. *Medium-term Budget 2003/04 to 2005/06*. Johannesburg.

City of Johannesburg and Monitor Company 2000. *Igoli 2010: Final Integrated Human Development Story and Final Integrated Story for the Mayoral Committee*. Johannesburg.

Demarmels, S. 2003. *Nachhaltigkeitsindikatoren für Santiago de Chile.* Praktikumsbericht. ETH Zürich.

Ducci, M.E. 2000. Santiago: territorios, anhelos y temores. Efectos sociales y espaciales de la expansión urbana. *EURE* Vol. XXVI, No. 79: 5–24. Santiago de Chile.

El-Shakhs, S. 1997. Towards Appropriate Urban Development Policy in Emerging Mega Cities in Africa. In C. Rakodi (Ed.) *The Urban Challenge in Africa – Growth and Management of its Large Cities.* Tokyo, New York, Paris: UNU Press: 497–526.

Gelderblom, D. and P. Kok 1994. *Urbanisation: South Africa's Challenge.* Volume 1: Dynamics. Pretoria: HSRC Publishers.

Gobierno de Chile 2002. *Census 2000/01*. Santiago.

Government of Botswana 1991. *Population and Housing Census*. Gaborone: Government Printer.

Government of Botswana 1996. *Living Conditions in Botswana: 1986 to 1994*. Central Statistics Office. Gaborone.

Government of Botswana 1997. *Population Projections 1991–2021*. Central Statistics Office. Gaborone.

Government of Botswana 1998. *Botswana National Settlement Policy*. Government paper No. 2 of 1998. Gaborone.

Government of Botswana 2000. *Botswana National Atlas*. Gaborone.

Government of Botswana 2002a. *Botswana Aids Impact Survey 2001*. Central Statistics Office. Gaborone.

Government of Botswana 2002b. *Statistical Bulletin March 2001.* Vol. 26, No. 1. Central Statistics Office. Gaborone.

Government of Botswana 2002c. *Gaborone City Council Recurrent Budget Summary 2001/2002.* Ministry of Local Government. Gaborone.

Government of Botswana Environmental Planning Programme (EPP) 2001. *Technical Report 9 – Gaborone City Council Institutions Dealing with Environment and their Linkages.* Gaborone: Government Printer.

Government of Botswana, HIES 1993/94. *Household Incomes and Economic Study*. Gaborone: Government Printer.

Graham, S. and S. Marvin 2001. *Splintering Urbanism: Networked Infrastructures, Technological Mobilities and the Urban Condition.* London: Routledge.

Harpham, T. and E. Werna 1996. Sustainable Urban Health in Developing Countries. *Habitat International*, Vol. 20, No. 3: 421–429.

Huchzermeyer, M. 2001. Housing for the Poor? Negotiated Housing Policy in South Africa. *Habitat International* (25): 303–331.

Kain, J.-H. 2003. *Sociotechnical Knowledge: An Operationalised Approach to Localised Infrastructure Planning and Sustainable Urban Development.* Dissertation. Department of Built Environment and Sustainable Development. Chalmers University of Technology. Gothenburg.

Keiner, M. and D. Salmerón 2003. Sustainable Urban Development in Gaborone, Botswana. DIMSUD working paper. *ETH e-Collection.* http://e-collection.ethbib.ethz.ch/cgi-bin/show.pl?type=bericht&nr=308, retrieved on May 6, 2004.

Lozano, E. 1990. *Community Design and the Culture of Cities.* Cambridge: Cambridge University Press.

Mahanyele, S. 2002. Minister of Housing's keynote address at Pre-WSSD conference: "Towards Sustainable Built Environment". University of the Witwatersrand, Johannesburg. August 2002.

Malbrán, H. 1998. *Desarollo Económico, Distribución Urbana y Motorización: Desafios para el Sistema de Transporte Urbano de Santiago.* Santiago: SECTRA.

McKenzie, E. 1984. *Privatopia: Homeowner Associations and the Rise of Residential Private Government.* New Haven CT: Yale University Press.

Molebatsi, C.O. 1996. Towards a Sustainable City: Gaborone, Botswana. *Ambio* Vol. 25, No. 2. Royal Swedish Academy of Sciences. Stockholm: 126–133.

Mosha, A.C. 1999. *Urban Poverty, the Case of Gaborone, Botswana.* A UNCHS/Habitat/ Ford Foundation Report. Gaborone.

Mosha, A.C. 2002. *Local Democracy and Decentralization in Botswana.* A UNCHS Report. Gaborone.

PNUMA-IEUT 2003. *Informe Geo Ciudades Santiago de Chile 2003.* Santiago.

Presidential Task Group for a Long Term Vision for Botswana 1997. *A Framework for a Long Term Vision for Botswana.* Gaborone.

Presidential Task Group for a Long Term Vision for Botswana 1997. *Vision 2016 – Towards Prosperity for All.* Gaborone.

Programa Chile Sustentable 1999. *Por un Chile Sustentable.* Propuesta ciudadana para el cambio, Santiago.

Rakodi, C. 1997a. Global Forces, Urban Change, and Urban Management in Africa. In C. Rakodi (Ed.) *The Urban Challenge in Africa – Growth and Management of its Large Cities.* Tokyo, New York, Paris: UNU Press: 17–73.

Rakodi, C. 1997b. Conclusions. In C. Rakodi (Ed.) *The Urban Challenge in Africa – Growth and Management of its Large Cities.* Tokyo, New York, Paris: UNU Press: 553–594.

Ravetz, J. 2000. *City-Region 2020.* London: Earthscan Publications Ltd.

Redefining Progress 2003. *Ecological Footprint.* http://www.redefiningprogress.org/programs/ sustainabilityindicators/ef/, retrieved March 27, 2003.

Republic of South Africa 2003a. *Municipal Profiles – City of Johannesburg.* Municipal Demarcation Board. http://www.demarcation.org.za/Munic_Profiles/New_layout/ index.html, retrieved on April 4, 2003.

Republic of South Africa 2003b. *Operational Plan for Comprehensive Treatment and Care for HIV and AIDS.*

Rogerson, C.M. 1997. Globalization or Informalization? African Urban Economies in the 1990s. In C. Rakodi (Ed.) *The Urban Challenge in Africa – Growth and Management of its Large Cities.* Tokyo, New York, Paris: UNU Press: 337–370.

Royston, L. and C. Ambert 2002. Going Against the Grain: Alternatives to Individual Ownership in South Africa. In G. Payne (Ed.) *Land, Rights and Innovation: Improving Tenure Security for the Urban Poor.* London: Intermediate Technology Development Group (ITDG) Publishing: 264–277.

Sassen, S. 2002. Locating Cities on Global Circuits. *Environment and Urbanization.* Vol. 14, No. 1: 13–30.

Silitshena, R.M.K. 1996. Urban Environmental Management and Issues in Africa South of the Sahara. In G. Benneh, W.B. Morgan and J.I. Uitto (Eds.) *Sustaining the Future – Eco-*

nomic, Social and Environmental Change in Sub-Saharan Africa. Tokyo, New York, Paris: UNU Press: 100–122.

Statistics South Africa (2003). *Census 1996 – City of Johannesburg.* Republic of South Africa. http://www.statssa.gov.za./SpecialProjects/Herman/wards/catb/Johannesburg.htm, retrieved on April 4, 2003.

Sustainable Cities 1992. Meeting Needs, Reducing Resource Use and Recycling, Re-Use and Reclamation. *Environment and Urbanization.* Vol. 4, No. 2: 3–8.

Sustainable Cities 2000. Towards more Pro-Poor Local Governments in Urban Areas. *Environment and Urbanization.* Vol. 12, No. 1: 3–11.

Sustainable Cities 2002. Sustainable Cities: Globalisation and Cities. *Environment and Urbanization.* Vol. 14, No. 1: 3–12.

Thale, T. 2002. *Johannesburg Plans to End Informal Settlements.* http://www.joburg.org.za/june_2002/housing.stm, retrieved on May 5, 2003.

Thrall, G. 1987. *Land Use and Urban Form: The Consumption Theory of Land Rent.* London: Methuen & Co. Ltd.

United Nations 1992. *Agenda 21.* http://www.un.org/esa/sustdev/documents/agenda21/english/agenda21toc.htm, retrieved on March 29, 2004.

United Nations Development Programme (UNDP) 2000. *South Africa: Transformation for Human Development.* Pretoria. UNDP.

United Nations Development Programme (UNDP) 2003. *Human Development Report 2003. Millennium Development Goals: a Compact Among Nations to End Human Poverty.* New York: Oxford University Press.

United Nations Human Settlements Programme (UN-HABITAT) 1998. *Global Urban Indicators Database 2 (1998 data).* http://www.unchs.org/programmes/guo/documents/1998.zip, retrieved on November 20, 2002.

Vestbro, D.U. and T. Almgren (Eds.) 1999. *How Should the Post-Apartheid City be Planned? Report from a Field Work in Port Elizabeth, South Africa.* Department of Architecture and Town Planning, KTH, Stockholm.

WCED The World Commission on Environment and Development 1987. *Our Common Future.* Oxford: Oxford University Press.

Wekwete, K.H. 1997. Urban Management: The Recent Experience. In C. Rakodi (Ed.) *The Urban Challenge in Africa – Growth and Management of its Large Cities.* Tokyo, New York, Paris: UNU Press: 527–552.

WHO World Health Organization 2003. *World Health Report 2003.* http://www.who.int/whr/2003/download/en/, retrieved on March 29, 2004.

Yousry, M. and Atta, T. 1997. The Challenge of Urban Growth in Cairo. In C. Rakodi (Ed.) *The Urban Challenge in Africa – Growth and Management of its Large Cities.* Tokyo, New York, Paris: UNU Press: 111–149.

Chapter 5

OPPORTUNITIES AND STRATEGIES FOR SUSTAINABLE URBAN DEVELOPMENT

Daniel K. Irurah[1], Björn Malbert[2], Aly Karam[1], Pål Castell[2], Marco Keiner[3], Branko I. Cavric[4]
[1]University of the Witwatersrand, [2]Chalmers University of Technology, [3]ETH Zurich, [4]University of Botswana

Abstract: Even though the global sustainable development paradigm has started to be translated into local visions for action in various developing countries and their respective cities, the process is extremely slow and fragmented. In particular, the absence of a comprehensive sustainability-based framework for national, regional and local planning indicates a lack of commitment to or incapacity for practical implementation beyond the political stage of signing global framework initiatives such as *Agenda 21* and *Habitat Agenda*. All three case study cities exhibit several pieces of policies, legislation and strategies and plans at national and local levels from which a comprehensive approach to sustainable urban development can evolve. Therefore, political will and institutional capacity at national and local levels are likely to remain the key determinant factors towards realizing integrated strategies in the case study cities as well as most other cities in developing countries.

Key words: Agenda 21, Habitat Agenda, backcasting, urban development visions and strategies, environment and resource management, governance and institutional capacity, economic growth and redistribution, equity and sociocultural development, privatization, public-private partnerships

M. Keiner et al. (eds.), From Understanding to Action, 115–156.

1. INTRODUCTION

1.1 Aim and Scope of the Chapter

The adoption of the sustainability vision and principles (as captured in key documents like *Agenda 21* and *Habitat Agenda)* into national and local planning is still limited and ineffective even though most countries are signatories to such protocols. This disjuncture between politics and practice in the processes of sustainable development is well manifested in the urban development of the case study cities. This chapter analyses the long-term development opportunities and strategies of the case study cities along themes of sustainable urban development.

In particular, a variety of strategies for sustainable urban development in the context of municipal planning and management will be explored as well as the key frameworks such as *Agenda 21* and *Habitat Agenda.* After introducing the concept and process of backcasting as a strategic planning approach for sustainable urban development, key strategies will be analyzed based on the case study findings grouped into four categories: environmental and resource management, land use management, economic growth and redistribution and governance and institutional capacity. Finally, the critical role of integrating sectoral strategies and plans as well as feedback provided by integrated monitoring, evaluation and reporting will be highlighted in the re-visioning and plan review processes.

1.2 Agenda 21, National Strategy for Sustainable Development (NSSD) and Local Agenda 21

As sustainable urban development depends on local action towards the collective global sustainable development vision, the related strategies are best analyzed within the framework set by several UN-brokered forums in the following documents:

- Chapter 7 in *Agenda 21* (UN 1992) on sustainable human settlements
- Chapter 28 in *Agenda 21* (UN 1992) on local authority initiatives
- Local Agenda 21 (LA21)
- *Habitat Agenda* (UNCHS 1996)

Before providing a detailed analysis of the strategies for sustainable urban development, it is necessary to provide an overview of the three aforementioned frameworks. However, it is important to note that the documents themselves are products of a backcasting approach in strategy analysis and formulation. In this regard, the next section will first give an overview on backcasting and demonstrate why sustainable development as referred to in *Agenda 21* and the *Habitat Agenda* should be seen as an outcome of backcasting at a global level to facilitating responsive action at the local level.

1.3 Overview on Backcasting

The sustainable development paradigm continues to be somewhat of an enigma, as reflected in current academic debates and political negotiation forums. This is possibly due to the fact that sustainable development emerged as the outcome of collective global backcasting, where the almost instantaneous capturing of an ideal future state led to the disregard for subsequent processes to search for concrete strategies and their implementation.

The practice of *backcasting* has emerged as a contrast to the conventional (but relatively passive) practice of *forecasting* as applied in conventional planning and management practice, especially in private sector organizations. Whereas forecasting is based on the extrapolation of historical and current trends into the future in order to predict and project future states and plan for them, backcasting starts with identifying desirable future states followed by the formulation of appropriate strategies to facilitate a transition to those states (Schmid et al. 1996).

In a detailed appraisal of backcasting and its application in the strategic analysis of organizations, Wang and Guild (1995) make the following observation:

> *"Backcasting firstly focuses on the range of alternative futures and their impacts, and then, determines the policy measures required to reach there. Rather than seeking accuracy-dependence of a predictive forecast as in conventional competitive analysis, backcasts intend to indicate the relative feasibility and implications of different end states a firm may have set for itself. In this sense, the futures become the 'cause' rather than the 'effect' of present planning and actions.*
>
> *[...] The major distinguishing characteristics of backcasting over forecasting is a concern, not with what futures are likely to happen, but with how desirable futures can be attained. It is thus explicitly normative, involving working backwards from a range of desired future end points to the present in order to determine the feasible steps need to be taken by a firm to reach that point [...]."*

Wang and Guild (1995), as shown in Table 5-1, capture the contrast between the two processes.

Table 5-1. Comparison of forecasting and backcasting as alternative approaches to futures studies (Source: Wang and Guild 1995)

	Forecasting	Backcasting
Question	What future is likely to happen?	How can desirable futures be attained?
Direction	Exploratory (opportunity-oriented) From present to future	Normative (goal-setting) From futures to present
Focus	Prediction and likelihood	Feasibility and choice
Execution	One time snapshot	Ongoing monitoring
Analysis	Extrapolation from historical data	Interpolation from goal setting, anchored on organizational competencies
Quality	Accuracy-dependent	Implication-oriented
Future(s)	Preceded by present assessment	Interpreted by present assessment
Result	Converge on the most likely future	Diverge to possible futures with respect to freedom of action

The backcasting framework is strongly aligned to the study and practice of sustainable (urban) development particularly because of its focus on desirable future states and the required process/strategies to ensure the transformation of present to future states. Backcasting, therefore, presents the "future" as a phenomenon to be strategically made and shaped through interventions transforming a present state to the desirable one. Figure 5-1 shows a diagrammatic representation of backcasting and its key elements.

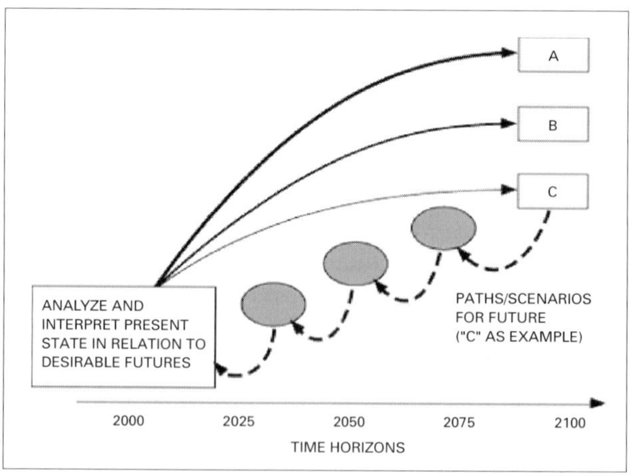

Figure 5-1. Key steps in the backcasting process (Source: Adapted from Wang and Guild 1995)

Backcasting practice often involves four key steps:
- Analysis of current state and trends (both desirable and undesirable ones)
- Envisioning desirable futures

- Formulating scenarios and strategies
- Monitoring and evaluation to guide review/refinement of actions/strategies

1.4 Sustainable Development as a Global Backcasting Outcome

The sustainable development paradigm emerged out of the realization that a projection of the current state and trends into the future clearly showed that human survival on earth was at tremendous risk from both environmental and socioeconomic threats arising from past and present patterns of production, consumption and the distribution of developmental benefits. In view of that undesirable future, the global search for a more desirable one led to the conceptualization of the sustainable development paradigm. *Our common future* (WCED 1987) consolidated the culminations of the first two backcasting stages as undertaken in a wide variety of previous processes (see for example *Limits to growth*, Meadows et al. 1972). Taking *Our common future* as embodying the first two stages of the global backcasting process, *Agenda 21* and the subsequent World Summit output (*Johannesburg plan of implementation*: UN 2002) constitutes the backbone of the third stage. The two scenario/strategy documents have been complemented with a variety of related stage-3 process/outcomes:

- *Millennium development goals* (UN 2000) which primarily focus on basic human needs like eradicating poverty, reducing illiteracy, enhancing access to basic services (shelter, water, sanitation etc) and improving health standards
- *Local Agenda 21* (LA21) which aims at facilitating action for sustainable development at the local authority level (based on Chapter 28 of *Agenda 21* – UN 1992)
- *Habitat Agenda* (UN 1996) which aims at facilitating action with regard to housing and settlements
- *Kyoto protocol on climate change* (UNFCC 1992) which aims at facilitating action with regard to mitigating global warming
- *Bio-diversity convention* (UNEP-CBD 2003) which would support and attempt to preserve bio-diversity globally

The crucial documents with regard to sustainable urban development are Chapter 7 of *Agenda 21* (sustainable human settlements), *Habitat Agenda* and *Local Agenda 21* (as per the framework of Chapter 28 of *Agenda 21*), as shown in Table 5-2.

Table 5-2. Key elements of three United Nations frameworks crucial to planning for sustainable urban development

Agenda 21, Chapter 7: Sustainable human settlements Key program areas	*Agenda 21*, Chapter 28: Local authority initiatives	*Habitat Agenda:* Key principles
– Providing basic shelter for all (affordable and environmentally sound) – Improving human settlement management – Promoting sustainable land use planning and management – Promoting integrated provision of infrastructure – Promoting sustainable energy and transport systems – Promoting planning and management in disaster prone areas – Promoting sustainable construction industry activities – Promoting human resource and capacity building	– *Local Agenda 21* for the community through community/stakeholder consultation/participation – Enhancing cooperation between municipalities – Enhancing exchange of information and experience – Enhance representation of women and youth in decision making, planning and implementation	– Equitable human settlements – Eradication of poverty – Sustainable development (need for economic growth, social development and environmental protection) – Quality of life for all (community and livable neighborhoods) – Family as basic unit of society – All people have rights – Partnerships – Solidarity with vulnerable and disadvantaged groups – Safeguarding interests of present and future generations – Human health and quality of life

The program areas envisaged under the three frameworks can be grouped into four major themes of governmental policy that also have been addressed in the case studies:

- Environmental and resource management
- Land use management
- Economic growth and redistribution
- Governance and institutional capacity

The key strategies for sustainable urban development will therefore be substantiated in Section 2 of this chapter along the four themes listed above. Each of the theme sections starts with a brief and general overview providing key objectives of required strategies and the broad mechanisms of such strategies. This is followed by illustrative content from the case study cities.

1.5 Backcasting in Planning for Urban Development: Vision 2016 in Botswana and Joburg 2030

Gaborone's planning in the 21st century will be happening in the context of Botswana's *Vision 2016*, which is the outcome of a national backcasting process. It is a multifaceted framework, which shows a nation aspiring to higher prosperity through enhanced competitiveness. Seven key strategic goals, which would facilitate this aspiration, were formulated as follows (Presidential Task Group 1997):

- Educated and informed nation (addressing the importance of education and training to meet national skill needs)
- Prosperous, productive and innovative nation (addresses the needs of a diversified economy especially through higher productivity in agriculture, industry, mining and services)
- Compassionate and caring nation (addresses the challenge of poverty eradication and equitable distribution of income)
- Safe and secure nation
- Open, democratic nation (addresses the need for decentralized democracy and political tolerance)
- Moral and tolerant nation (addresses the need for discussion and openness on issues such as gender, disability, cultures, religion etc)
- United and proud nation sharing common ideals, goals and symbols

It is envisaged that the seven strategic goals will constitute the primary reference for future planning, especially for the country's national development plans, as well as future revisions to existing plans. It will therefore have fundamental implications for urban and rural planning outcomes and processes. However, it is quite surprising that, being a development framework in the 21st century, *Vision 2016* goals have not been referenced to sustainable development, especially the need for resource and environmental conservation.

Through a similar process, Johannesburg's planning now happens in the context of a broader city vision, *Joburg 2030*, which was also formulated through a backcasting process. As further discussed in Section 7 and Chapter 7, Section 1.3, the city wants to expand its global and regional competitiveness in order to maintain and enhance its status as African world class city. As in *Vision 2016* in Botswana, *Joburg 2030* prioritizes economic growth as the key strategic goal with socioeconomic challenges as secondary priorities. Resource and environmental challenges and outcomes play no significant role in the vision.

Given the strong influence that the two visions will have on the future planning and development of their respective cities, it is unlikely that sustainable urban development will be comprehensively addressed in the next years. However, given the increasing local and global pressure to strive for sustainable urban development goals, it is possible that the two visions will be responsively revised before their planned time frames mature.

2. STRATEGIES FOR ENVIRONMENTAL AND RESOURCE MANAGEMENT

Environmental and resource threats are perceived very differently by stakeholders in developing countries as opposed to those in developed countries. This is particularly due to the pressing basic-needs challenge where enhancing standards of living takes first priority and the average ecofootprint of lifestyles and cities is still relatively low (Sustainable Cities 1992).

The awareness of the need to restrain the ecofootprint as standards of living improve is still somewhat limited. Consequently, the most common understanding of the environmental and resource constraint is where it directly impacts general health and services. This creates the view that there are local environmental issues, which require immediate attention, and global issues, which can only be significantly addressed once basic needs have been effectively met. This resonates with the *Agenda 21* principle of differentiated responsibilities between developed and developing countries. However, it contradicts the global expectation that developing countries must aim to leapfrog into clean technologies (and avoid taking the unsustainable path historically taken by developed countries) as they engage with development and fulfilling basic needs.

The local environmental/resource issues likely to receive high priority include the following (Sustainable Cities 1992):

- Supply of land for housing with regard to equity and poverty reduction
- Waste management as a response to environmental health
- Air quality as a response to environmental health
- Water quality and scarcity (in periods of drought or floods)
- Sanitation and its threat to environmental health especially through water impacts and insect-borne/parasitic diseases
- Absence of green and open spaces as a recreational constraint

The global issues likely to be ignored or receive inadequate attention include the following:

- Global warming and climate change
- Conservation of bio-diversity
- Clean technologies in production (less inputs, more reuse/recycle, application of industrial ecology principles and eco-business approaches, etc)
- Overproduction and overconsumption (especially by the privileged minority residents)
- Water scarcity as a global threat
- Deforestation and desertification

One example that demonstrates this split approach is where delivery of water services is highly prioritized as a basic need (potable water, sanitation and equity issues), but water conservation strategies are rarely integrated in such programs. Similarly, one finds concerted efforts to improve waste disposal services (collection and disposal in landfills), but rarely do such efforts include integrated waste management strategies (reduce, reuse and recycle before disposal). Even where economic benefits and opportunities exist through such integrated approaches (e.g. job creation), the synergy is rarely recognized or exploited systematically.

2.1 Strategies for Energy Efficiency and Clean Air

Although the decoupling of energy and development has become a strong sustainability objective in developed countries (Azar et al. 2002), they still struggle with the need to access more affordable and cleaner energy services. It is therefore expected that increasing levels of per capita energy consumption and related impacts will accompany their development.

In developing country cities, environmental issues relating to energy take two dimensions. The first one relates to air quality due to local air pollution at the household, neighborhood and urban scales. The second one relates to the global issue of fossil fuels and global warming. In general, the former, representing the more immediate and pressing need, takes priority over the latter.

For example, the use of combustible fuels at the domestic scale (wood and wood products, coal, paraffin, etc.), especially by poor households, generates particulate and gaseous emissions, which contribute to indoor and outdoor air pollution. High levels of such pollution have been linked to the prevalence of respiratory diseases in specific zones of such cities. Promoting the use of cleaner energy services and renewable energy for space heating, cooking and water heating is therefore likely to be driven by environmental health and related costs rather than by emissions reduction for greenhouse gases. Passive thermal control, solar water heating and access to electricity or petroleum/natural gas for cooking are examples of strategies promoting better environmental health, which have minimal effect on the mitigation of global warming (Irurah 2000). Botswana has very good conditions to promote photovoltaic plants, since there is a lot of sparely used desert land where this equipment could be installed. Solar energy can also be used for water heating. If the country changed its energy resources policy to renewable ones, it would be a big step towards sustainable development.

Air pollution related to transport is also likely to be prioritized in this local context approach. The challenge of smog, low-level ozone and particulate emissions (especially lead from leaded fuels) is likely to attract more attention than the challenge of global warming impacts (Republic of South

Africa 2003). Consequently, energy-related strategies tend to fall in three broad categories, prioritized as follows (Irurah 2000, Du Plessis et al. 2003):

- Increasing access to conventional energy services, especially for the poor
- Minimizing and monitoring localized air pollution impacts
- Mitigation of greenhouse gas emissions (GHG)

Enhancing access to conventional energy services - Increasing access for poor households and small/informal businesses to conventional energy services as a means towards improving standard of living is one of the top developmental priorities in developing countries and is often closely tied to housing delivery. For sustainable development to be enhanced, strategies under this category should integrate energy efficiency/conservation and renewable energy technologies, with a long-term goal of evolving towards CO_2-neutral and factor-four technologies (see Von Weizsäcker et al. 1997 for best practice examples).

Minimizing and monitoring localized air pollution impacts - Electrification, the use of cleaner energies like petroleum and natural gas and renewable energies, like solar energy, are some of the strategies under this category. Minimisation can be further enhanced through urban greening (for CO_2 sinks) as well as responsive urban development in relation to meteorological dynamics such as temperature inversions and prevailing winds. Although the challenge of air pollution is related to land use, and especially urban sprawl and social segregation in certain zones, it is likely to get worse through high densities and compact urban forms. Consequently, pursuing compact city principles without integrating cleaner energy technologies can only magnify the localization and escalation of such air pollution impacts (Du Plessis et al. 2003). Mitigation through the *polluter-pays principle* is usually not easily enforceable in developing country cities for a variety of reasons such as poverty, administrative burden and inadequate enforcement-capacity (financial and skills/competencies).

Given the serious impacts of air pollution in Santiago, regional and municipal authorities recognize the need for long-term solutions. Therefore, it is clear that local air pollution impacts will attract faster legislative and policy response compared to global challenges of greenhouse gas reduction to mitigate climate change. Policies and strategies for reducing local air pollution due to vehicular and industrial emissions may therefore contribute to faster changes of values, lifestyles and consumption in Santiago. This would, in turn, contribute to the mitigation of greenhouse gas emissions and related climatic impacts as one of the secondary outcomes.

Environmental monitoring and reporting are key elements for coping with air pollution. The high local impacts related to air pollution in Santiago have inevitably prioritized air quality monitoring and reporting in a way which facilitates decision-making on a daily basis as well as for long-term policy planning for land use, transportation and industrial technologies. In the last twenty years, such monitoring has been systematically implemented through fixed and mobile stations operated by the *MACAM Network* (Official Network of Automatic Monitoring of Air and Meteorological Quality). The monitoring generates data under two major categories: fixed and mobile sources. Fixed source emissions are from spatially fixed emitters such as heating boilers, smelter ovens, industrial boilers and incinerators. Mobile source emissions are predominately transport-related.

The monitored data are systematically fed into an air quality modeling system, which simulates air quality levels in successive 24-hour periods. In order to facilitate emergency decision-making, the model applies three categories of pollution levels: alert, pre-emergency and environmental emergency. In case of emergencies, decisions can range from imposing vehicular restrictions to suspending industrial activity or school attendance.

Mitigation of greenhouse gas emissions - Energy conservation and renewable energy strategies for mitigating greenhouse gas emissions may probably not take root in developing country cities until the standards of living have improved significantly for stakeholders to acknowledge and prioritize the need. It is likely that increasing costs of conventional energy from fossil fuels will trigger the required strategies rather than the planning objective of mitigating GHG emissions. With regard to urban form, local authorities are likely to pursue compact city forms and high densities to respond to impacts on the relative costs of infrastructure and services, such as transportation, rather than to the global challenge of global warming.

Thus, developing country cities are faced with a strategy dilemma. Whereas it could be advisable to pursue a full-cost approach to pricing all energy options (including the comprehensive range of externalities for each option) and give incentives for efficiency/conservation and renewable energy technologies for local and global sustainability, such an approach would continue to marginalize the poor, denying them access to reliable and cleaner energy services unless some form of subsidy is integrated into such strategies in the short to mid-term (Sustainable Cities 1992, Irurah 2000).

2.2 Waste Management Strategies

Even though the awareness of the "3R" approach to integrated waste management (reduce, reuse, recycle before disposal) is becoming widespread in most developing country cities, the primary policy concern is still the high level of waste disposal that endangers environmental health. This is because very few local authorities have reached operational efficiency in collecting and disposing waste reliably for the entire municipality. Whereas the informal settlements and poorer neighborhoods are the most affected, collection and disposal service even in middle to high-income suburbs and commercial and industrial areas is also not always consistent.

However, even in such a context, poverty and unemployment has led to the unregulated scavenging of waste for valuables, ranging from food and other reusable valuables to recyclable materials like paper, cardboard, timber, steel, glass, aluminum cans etc. In fact, for poor urban households, there is a direct connection between recycling and energy, because most of the combustible fuels used for cooking and heating comes from salvaged wood, cardboard or paper. Similarly, most of the materials and components used in the construction of informal settlements are often salvaged at various stages of the waste stream of construction and non-construction materials, especially cardboard, wooden planks and corrugated iron sheets (Sustainable Cities 1992).

The emergence of recycling as a formal-sector linked activity, especially for paper and cardboard, glass and steel, has opened up opportunities for local economic development (LED) programs often supported by public and private sector stakeholders as well as communities and non-governmental organizations. However, the range of recycled waste still remains narrow, as only a few industrial sectors have adopted systematic industrial ecology principles.

Therefore, it is clear that the effective collection and disposal of waste will remain a priority for cities in developing countries. Integrated waste management will only grow rather slowly especially in response to transforming formal sector industries with industrial ecology strategies. Similarly, opportunities for enhancing LED (especially in relation to job creation and business opportunities), rather than global concerns for resource conservation, are likely to prevail as the key drivers of integrated waste management (Sustainable Cities 1992).

2.3 Water Provision, Conservation and Sanitation Strategies

The primary strategic focus for water services in developing country cities is more equitable access, especially for poor households. Although capital subsidies for related infrastructure networks are often provided by the national government (often linked to housing subsidies), such subsidies do not cover the operational or consumption costs incurred by local authorities or their agencies. Consequently, local authorities are often faced with the burden of offering services without receiving payments from poor households. The provision for a minimum level of service at no cost to the poor households can be implemented, possibly by increasing tariffs at higher levels of consumption as a cross-subsidization strategy. The targeting of such tariffs to benefit overall water efficiency and conservation could also be pursued (Sustainable Cities 1992).

As noted above, the lack of water efficiency and conservation measures in the water services program is a clear indication of the low priority, which most governments (national and local) give to the global water scarcity challenge. Measures such as integrating alternative sanitation systems, reusing grey water, especially for urban agriculture or flushing toilets, and water saving devices, such as dual flush toilets, aerated taps and low-flow showerheads, are rarely integrated or evaluated at the initial stages of major water distribution projects or in the design and construction of buildings (Suzenet et al. 2002). Similar initiatives targeted at high-level consumers (middle to high income households as well as commercial and industrial consumers) are either lacking or implemented in an unsystematic manner. In addition, the absence of appropriate legislation and policies in this regard inhibits initiatives by relevant stakeholders.

Water conservation strategies within the municipal distribution network (pressure regulation and minimizing unaccounted for water loss through early detection and response to burst and leaks) are also crucial as most municipalities waste large portions of water daily. Similarly, the regular monitoring and upkeep of the water network is crucial to minimize losses and costs related to aging pipes, valves and pumps. In fact, through water conservation and an efficient distribution network (pressure regulation and efficient pumps), municipalities can achieve energy savings and lower operational costs (Suzenet et al. 2002).

2.4 Overproduction, Overconsumption and Related Resource/Environmental Conservation Strategies

At policy level, very few developing countries prioritize overproduction and overconsumption as serious environmental threats. This is based on the perception that only an extremely small proportion of the population earns income levels that allow overconsumption. Hence, their low per capita ecofootprint is unlikely to warrant serious attention when compared to the challenge of meeting basic needs of the poor majority.

However, the above scenario is rather simplistic for two reasons:

- As development takes effect and standards of living improve, the increased consumption and production will be driven by antiquated values, values which currently drive overconsumption and overproduction in developed countries
- A significant proportion of the production in developing country cities is targeted for external markets. Consequently, large external markets for local resources can contribute to overproduction locally with overconsumption happening elsewhere (Sustainable Cities 1992).

Since the overall transformation of an economy constitutes the fundamental framework for appropriate strategies under this challenge, with firms and households as the key role players, municipal authorities would have to act in partnership with the national government for the formulation, implementation and enforcement of appropriate policies and legislation (see Section 5.2). The evolution of alternative value systems centered outside of the conspicuous consumption of goods and services as well as an increased application of clean technologies (especially based on principles of industrial ecology and life-cycle costing) are some of the overriding requirements for strategies under this challenge (Sustainable Cities 1992).

3. STRATEGIES FOR LAND USE AND CONSERVATION OF BIODIVERSITY

Strategies for land use often stem from a land use management vision to manage the spatial growth of a city in relation to the economic and sociocultural demands as well as environmental conservation objectives. The key principles of a land management strategy include the following:

- Spatial efficiency through urban integration and appropriate densities (doing more with less: this has implications on the requisite capital investment for supportive infrastructure)
- Minimize establishment and transaction costs for firms and households (minimizing travel distances/costs/time and traffic congestion, mitigating high property costs and rents, etc.)
- Equity in terms of access and tenure for basic shelter and services for the poor and vulnerable (women, aged, disabled, refugees and those impacted by HIV/AIDS)
- Open space system allowing for a spectrum of needs (recreation, natural conservation and biodiversity, environmentally sensitive areas like catchments and flood zones as well as high potential areas for agriculture and cultural heritage)

In contrast to the conventional and often passive spatial planning and zoning practice, land use management strategies for sustainable urban development aim to reinforce the municipal authorities as pro-active agents in guiding and facilitating urban growth and not merely reacting to or regulating spontaneous development initiatives by private developers. Consequently, such strategies are characterized by a wide variety of tools and mechanisms:

- Declaration of growth boundaries (even within a given municipal area, growth boundaries would be different from the administrative boundaries of a municipality, primarily because development would only be encouraged on areas enjoying sufficient levels of infrastructure and services while leaving the rest as urban open space or for agricultural use).
- Concentration of infrastructure and services within areas intended for intensive development (nodes and corridors).
- Building permits aligned with development objectives as well as social integration and mixed-use principles.
- Control of land invasions and illegal settlements through integrated housing development, especially by applying housing subsidies towards the realization of overall desirable urban outcomes to avoid spatial segregation and sprawl.
- Formalization and upgrading of informal settlements, integrating amenities and services as well as decreasing the density of overcrowded areas.
- Incentives for the revitalization of established areas faced by blight and decline (primarily because of the migration of businesses to newer nodes in the suburbs).

In order to monitor land use impacts of urban and economic growth, the responsible authority requires a wealth of spatial information in different formats. When this is coupled with the need for spatial modeling in order to generate scenarios for decision-making, the role of a comprehensive geographic information system (GIS) becomes evident. The use of GIS requires strong institutional capacity in terms of financial resources and skills, not only for managing the system but also for innovating and modeling alternative strategies or revisions and refinements based on insights from monitoring and evaluation processes or from best practice examples.

In conclusion, it should be noted that it is not possible to prescribe sustainable urban form and land use management beyond the statement of principles and primary objectives. This is mainly because land use is ultimately a spatial manifestation of the socioeconomic and environmental strategies and priorities which stakeholders have evolved and agreed upon. Consequently, although one can talk about thresholds for certain outcomes (densities required for the viability of public transport or other amenities like schools and clinics), it is not possible to prescribe a specific density as the sustainable one irrespective of all those other considerations.

3.1 National Policies and Strategies on Land Use and Settlements in Botswana

Botswana has already initiated national strategies and policies in spatial planning and land use management. For example, the *National Settlement Policy* (NSP) aims at countering the developmental polarization between urban and rural areas by ensuring more equitable distribution of trained manpower and financial support to settlements, according to their specific roles within a more equitable national development agenda (Government of Botswana 1998). Moreover, if developed as intended, a proposed *National Physical Plan* (see also Chapter 7, Section 1.6) will facilitate a system of urban and rural settlements developed in relation to the location of resources, to promote a more balanced economic growth throughout the country and permit the provision of infrastructure and other facilities on an economic basis. Such a planning instrument could also promote the diversification of urban centers in addition to Gaborone and Francistown, thus providing more alternatives for the absorption of the migrant population and avoiding the problems arising from excessive concentration in existing major urban areas (Government of Botswana 1998).

Finally, *Regional Master Planning* in each of the four planning regions (a requirement of the *National Settlement Policy*) will provide an overall framework for all district settlement and local plans, setting general directions for district and settlement development, infrastructure and social service provisions, economic/financial requirements and appropriate land utilization. In this way, it may become possible to provide a framework for the spatial distribution of resources and to facilitate the creation of incentives for increased production through investment by both private and public sectors.

At the scale of urban settlements, the future *National Physical Plan* will also require the formulation of *Urban Settlement Development Plans* (USDP), which should be aligned to the *Regional Master Plans*. The USDP must address environmental, socioeconomic and urban management objectives, programs and projects. The overall nesting of national and municipal economic and spatial planning is shown in Figure 5-2. National and Urban Development Plans are five-year economic plans that are used to prioritize the projects of the Government and the Councils. The plans specific for Gaborone are treated in Chapter 7, Section 1.6.

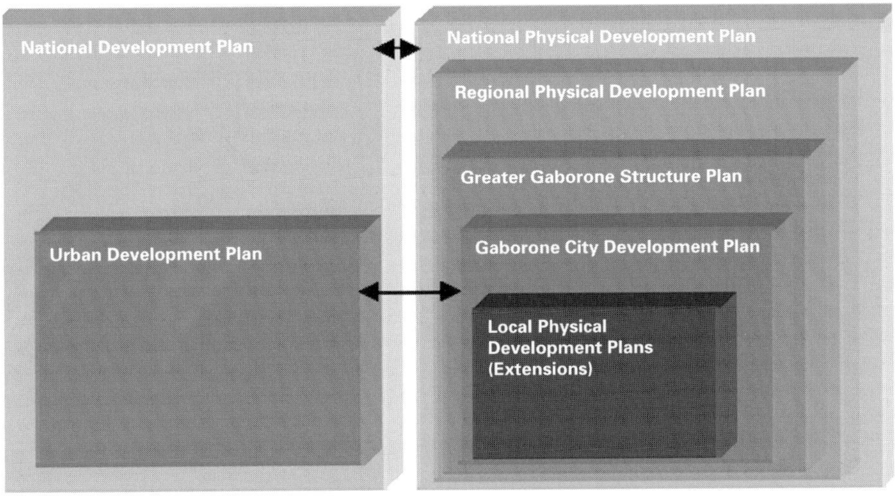

Figure 5-2. The nesting of municipal versus national spatial and economic policies and plans in Botswana (Source: Thalmann 2002)

3.2 The Spatial Development Framework for Johannesburg

The *Spatial Development Framework* (SDF) is one of the critical components of Johannesburg's *Integrated Development Plan* (IDP), legislated under the *Municipal Systems Act* of 2000. The primary objective of the SDF is to serve as the spatial translation of the city's socioeconomic and environmental development objectives with urban integration, enhanced accessibility and city efficiency as the targeted outcomes. The SDF is aligned to the provincial spatial development framework of Gauteng Province, in which Johannesburg is located. It is based on four key building blocks as shown in Figure 5-3:

- Spatial plan
- Investment framework
- Instruments
- Local plans

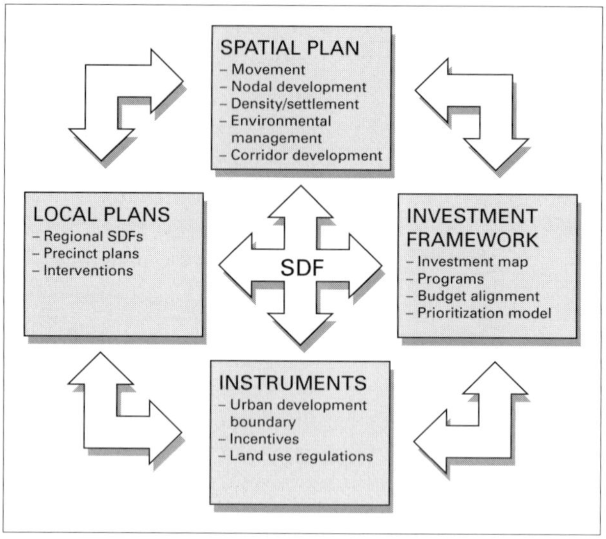

Figure 5-3. Key elements of the Spatial Development Framework of Johannesburg (Source: Redrawn from City of Johannesburg 2003a)

Each of the blocks constitutes various elements dynamically linked in order to maximize impact (City of Johannesburg 2003a). The spatial plan defines and elaborates the strategic spatial elements and policies of the SDF. These include a movement system, nodal development, settlement densification and environmental management (including the *Johannesburg Open Space System* – JMOSS), corridor development and development guidelines). The investment framework provides a mechanism for investment scenario modeling and prioritization in relation to long-term and annual priorities.

The instruments provide the mechanisms and tools for the realization of the SDF. These include the urban development boundary, incentives and land use regulations. Finally, the local plans constitute the interpretations of the SDF in terms of regional SDFs, precinct plans and key intervention programs. The aforementioned building blocks and the elements of the SDF have all evolved through several annual reviews of the SDF and IDP.

3.3 Land Use Planning as a Resource and Environmental Conservation Strategy in Santiago

In Chile, the planning instrument for addressing the urban sprawl issue is the *Santiago Metropolitan Master Plan* (PRMS; see Chapter 7, Section 1.6).

The PRMS (MINVU 1994) differentiates between "Urban Areas" and "Urbanizable Areas." The former correspond to territories that were consolidated when the plan was formulated in 1994, for which regulations must be implemented in those areas to densify and intensify land use. Urbanizable Areas extend around the consolidated urban area, where the city's future growth is anticipated. The sum of the two zones, is delimited by the "Urban Extension Boundary." Areas excluded from urban development lie outside of that boundary. Figure 5-4 shows the location of urbanizable (striped) and restricted (checked) areas.

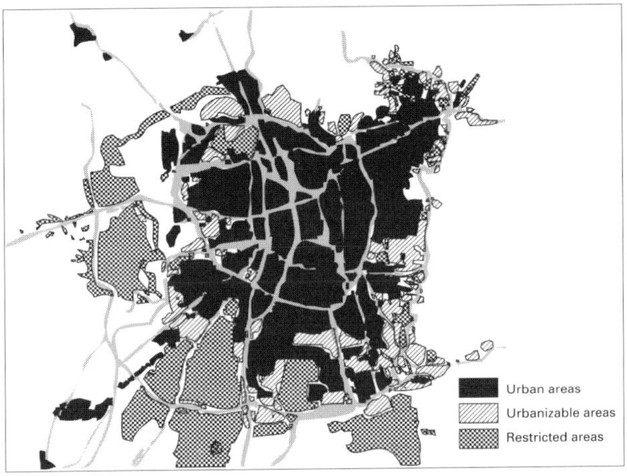

Figure 5-4. Land use planning as an environmental strategy in Santiago (Source: Redrawn from MINVU 1994)

Consequently, the PRMS defines a boundary beyond which no kind of urban occupation should occur, unless it is intended for agriculture, agribusiness activities, or the protection of the natural environment. The urban extension boundary could therefore contribute to trend stabilization in Santiago's urban sprawl.

4. ECONOMIC GROWTH AND REDISTRIBUTION

The overall objective of economic growth in the context of sustainable development is to increase overall levels of GGP (Gross Geographic Product) while ensuring equitable distribution of the benefits. One of the subobjectives should be to mitigate the tendency towards concentration of wealth both in periods of growth as well as stagnation or decline (Sustainable Cities 2002).

Enhancing and sustaining economic growth in a globalized world is one of the key challenges for a systematic backcasting approach, especially with a view to transforming the legacy of weak comparative advantages in global or regional competitiveness and attractiveness. Countries and cities, which aim to sustain their economic growth, are also likely to transform their competitive advantage in response to global transformations and needs.

One of the key shortcomings of contemporary planning in developing country states and cities is the approach of projecting forwards from an extremely weak legacy and at very short intervals of about 5 to 10 years (often linked to electoral periods), which often reinforces the business-as-usual scenario. The understanding of a need to reposition in the context of a globalized and competitive world is often limited and visions involving periods of one or two decades is extremely rare (Sustainable Cities 2002).

A key element in global competitiveness is lowering establishment and transaction costs for organizations as well as minimizing risks to investment in order to ensure competitive returns for firms and investors. Lowering establishment and transaction costs closely relates to city efficiency in terms of its infrastructural networks and services as well as the pool of human resources. Investment risks are mainly related to direct threats, such as crime and civil war, as well as political instabilities, unpredictable/unstable policy and legislation, levels of taxation on incomes and profits and the frequency of civil unrest or strikes (Sassen 2002).

It is clear that enhancing a city's comparative advantages and economic growth is strongly linked to the national and regional economic as well as political environment. The major question arising from the issues above would be, "Can cities really make a difference in investment decisions of internal or external investors?" With regard to external investors, one would actually accept that there is a hierarchy of roles in view of a globalized world. For example, the emergence of major economic cooperation and trading blocs reflects the need to go beyond cities and states when targeting global investments and trade opportunities. In both developing and developed countries, the idea of consolidating large markets and harmonizing the operational environment for businesses has led some countries to abandon the competitive approach for cooperation.

Therefore, global and regional cooperation frameworks constitute the first level of strategies for economic growth within a city. However, such frameworks do not necessarily confer benefits equitably either among the states or the respective cities. Consequently, there is always a need for such blocs to define how equity within the bloc will be addressed, especially because automatic reliance on comparative advantages or attractiveness could easily lead to the concentration of benefits within one state or city (Sustainable Cities 2002).

The logic of regional economic blocs versus single states is closely related to that of state versus individual cities in that through policy, legislation and programs (both economic and otherwise), the state plays a major role in shaping attractiveness for investments within the country in general. For example, the balance in national budgets between capital development (especially infrastructure development like power, roads and telecommunications) and operational costs of the state would significantly influence resources available for infrastructure networks within a city as well as interconnections between cities and their hinterlands.

A city's comparative advantage can therefore be enhanced or undermined by national policies and programs. It is critical to recognize this as the second level of strategies. In particular, policies relating to crime control and the judicial branch, education, transportation, telecommunications and energy have major consequences on the city's attractiveness nationally, regionally and globally. The inability of states to provide a reliable sociopolitical environment has been viewed as one of the major investment risks that developing countries must overcome regionally and globally. The African Union (AU) and NEPAD initiatives in Africa are some of the approaches towards addressing this challenge at a continental scale. In a democratic environment, cities need to strongly support their agenda within national and regional policy making structures.

Taking the global, regional and national contexts as given, what are the key strategy options for cities at the municipal level? These can be grouped into five key categories as follows (Sustainable Cities 1992):

- Strong alignment of national, regional and global initiatives. Such alignment would be effected through responsive municipal programs, bylaws (formulation and enforcement), budgeting, and overall management
- Taking care of the fundamentals, especially in regard to municipal level infrastructure and services, to facilitate efficiency in production, distribution and marketing at all levels
- Providing investment incentives, including lower costs for levies and rates on business premises, cost-competitiveness on inputs like energy and water and support for innovation and the incubation of new lines of business in sectors identified as strong clusters for the municipality
- Facilitating effective partnerships especially aimed at optimizing synergies between complementary investment opportunities among business clusters
- Marketing the city to different stakeholders. This translates to initiatives in city image, branding and marketing the city at municipal, regional, national and global levels. Although branding and marketing the city for tourism has become common, targeted marketing for other so-

cioeconomic investment has not been systematic in developing country cities.

Since the formulation and implementation of these strategies require highly responsive governance, the institutional capacity of municipalities has a strong bearing on their attractiveness and hence on economic growth. This is discussed in detail under Section 5.

4.1 Employment and Equity

With the decoupling of economic growth and employment, with jobless growth as one of the manifestations (see Chapter 4, Section 3.2), the pursuit of growth alone will not automatically address the poverty challenge, especially in cities where literacy and skills level are still low or in decline. In this regard, the role of *Local Economic Development* (LED) and *Small, Micro and Medium Enterprises* (SMMEs) becomes critical. Growth driven by LED and SMMEs tends to generate more jobs per unit of investment because the decoupling effect is often minimized at smaller scales.

In view of the above, cities must not only face the challenge of attracting and retaining externally sourced investments, but also create a variety of levels of opportunities in which SMMEs can incubate and grow. Where cities fail to systematically manage this challenge, the emergence of a dual economy becomes inevitable. This is characterized by a relatively small proportion of the population actively engaged within a stagnant or shrinking formal economy (often associated with relatively high remunerations), with the majority in the insecure and poorly remunerated informal sector or SMME jobs (Sustainable Cities 1992).

4.2 LED as an Economic Strategy

Local Economic Development (LED) planning has emerged in the past decade with the realization that unemployment and poverty conditions are getting worse even in cities where economic growth has been increasing. The three key attributes of LED planning are as follows (Sustainable Cities 2000):

- Mainly practiced as a poverty alleviation strategy. Consequently, it tends to be targeted to city areas experiencing high levels of unemployment and poverty.
- Strongly focused on local opportunities in terms of markets and production. However, linkage to the larger economic system can be one of the targeted outcomes.

– Strongly targeted at small-scale production (SMMEs, individuals and households) and distribution.

There are two approaches to LED programs and initiatives. The first one aims at *reinforcing and facilitating growth* in already existing businesses, which have been constrained due to entrepreneurial, capital or market limitations. Such businesses could be in both the formal or informal sectors. The approach in this case is based on capacity building by facilitating access to additional capital, skills and other inputs like land or equipment.

The second approach involves the full entrepreneurial cycle where both the *opportunity and production processes* are developed from the beginning with certain beneficiaries in mind. Since such initiatives are externally driven, they tend to take the formal rather than informal business structure. Although the targeted communities are a critical component of the LED program, the drivers on the facilitation side could consist of a variety of role players (private sector, NGOs, municipality or central government).

Besides addressing unemployment and poverty, LED programs could also have an *environmental dimension* where environmental solutions constitute the core of the opportunity. A common example is waste management through separation, reuse and recycling before disposal (see Section 2.2). Also, energy services companies have been successfully established (for the management of clean energy supply and distribution, energy efficiency and renewable energy technologies) in various communities in developing country cities.

4.3 Addressing Poverty through Job Creation Programs and Micro Credit Allowances

Poverty alleviation is a key issue for urban development in developing countries. Without meeting the needs of the poorest and disadvantaged, sustainable development cannot take place. Jobless growth, as discussed in Chapter 4, Section 3.2, increases the socioeconomic segregation among citizens and does not offer access from unemployment or from the informal sector to formal jobs. Thus, job creation programs are often planned as targeted initiatives within the formal sector of the economy and especially within the government's infrastructure development projects. Such projects are to be implemented through labor intensive methods with the overall objective of enhancing job opportunities as well as the participation of SMMEs with previously disadvantaged groups according to gender, race, disability, etc as the main beneficiaries (affirmative action).

Under such programs (covering issues such as housing, water, sanitation, electrification, etc), organizations implement procurement policies identifying job creation as one of the targeted outcomes. Implementing agencies are awarded projects on a competitive basis with technical, cost and job or SMME participation opportunities as the key criteria. The public sector may also be involved in complementary initiatives, such as raising awareness and providing training and capacity for both targeted beneficiaries and implementing organizations (especially the professionals in charge).

Given the declining role of the public sector as a major contributor to investment and economic growth, it is unlikely that job creation and LED programs can effectively overcome the impacts of structural changes experienced by modern economies contributing to the trends of increasing unemployment and poverty even during periods of economic growth (Sassen 2002, Sustainable Cities 2000)

It is equally unlikely that any government (local or national) can sustain the multiplicity of initiatives required for different communities, especially where such initiatives are not inherently integrated into the formal sector economy. Initiating such projects and then terminating them before they have matured to full-fledged opportunities could lead to further socioeconomic distress among communities. In conclusion, it is clear that a major innovation in distributive strategies and mechanisms is required politically and economically, if the poverty and inequality challenge is to be adequately addressed in the context of decoupling growth and employment in developing country cities.

A successfully proven strategy to enhance income generation of the poor is the granting of small bank credits. Micro credit facilities allow for small investments for the production or trade of goods. Micro credit is the extension of small loans, often under US$ 100, to entrepreneurs too poor to qualify for traditional bank loans. It has proven to be an effective and popular measure in the ongoing struggle against poverty. In developing countries, micro credits are given by NGOs, social organizations and others to individuals (preferably women) or to cooperatives. If interest rates are low and the payback time limits are long, the credit users have the chance to gain benefits from their investment, even founding small enterprises. By this, the vicious circle of unemployment/poverty/crime can be broken and additional income means better tax yield for the state, thus increasing the national GDP.

A well-known model of a successful micro-enterprise strategy is that of the *Grameen Bank*. Founded in 1979 in Bangladesh, the Grameen Bank has been helping to alleviate poverty and to promote economic development (Higinbotham Holcombe 1995, Yunas 1999). Until 2003, over 2.4 million families benefited with over US$ 3.7 billion in micro credits or loans (Grameen Foundation USA 2003). However, their success has been limited. Even with micro credits, most borrowers remain at the poverty line and many small enterprises stay small or disappear due to a lack of knowledge about market access and business management. Thus, basic education and economy skills training is necessary to make micro credits a true success story.

5. GOVERNANCE AND CITY MANAGEMENT

The challenge of sustainable urban development crosscuts through all sectors and levels of government (see Chapter 7, Section 1.4). The building of relevant institutional capacity is a priority issue in all cities of the world and thus a problem area where cities can cooperate and learn from each other within and between regions of the world. Also, strong institutional capacity may be a barrier to change if not redesigned according to the vision of sustainable development (see Fudge and Rowe 2000 for the Swedish context). The case studies of the DIMSUD project indicates at least three critical items to be addressed when forming strong political leadership for sustainable development:

- The coordination of legislation and public sector administration at all levels in a way that supports transparency, efficiency and comprehensive approaches.
- The distribution of responsibilities, power and resources between levels of government in accordance to what actually can be planned and achieved at each level, leaving the choices as close as possible to those who are supposed to implement the measures required.
- The ability of the government system to manage partnership and participation arrangements in a way that promotes fair development and democracy for all.

5.1 The Role of the State versus the City in Sustainable Urban Development Strategies

Even though decentralization in planning and decision-making is crucial for sustainable urban development, the role of the state in steering overall policies is critical. For example, it would be easier for the city to implement effective land use, environmental and resource management, economic growth or equity strategies where clear and enforceable national legislation, regulations and procedures are already in place. This is one of the contexts where national policies on issues like land use, energy, water and environmental impact assessments become a major opportunity for sustainable urban development strategies as the city needs only to translate them into their planning and development control system.

Through a review of the existing opportunities and strategies for sustainable urban development in the case study cities, minimum policy, legislative and institutional requirements have started to emerge. In particular, the hierarchical nesting of national policies and legislation and their respective translation into regional and municipal level strategies and plans is evident in all the three cities.

In relation to institutional frameworks, the case study cities also provide a variety of approaches towards coherence and streamlining of institutional structures in relation to formulating instruments to meet sustainability challenges. This recognition of the interdependency between local, regional and national policies is possibly one of the most critical opportunities as it generates the political will to ensure coherence and streamlining of policies and plans in different spheres. The slow progress towards achieving stabilization or the reversal of negative trends is possibly due to inadequate capacity to translate policy and legislation coherence envisaged on paper into the reality of concrete initiatives.

5.2 MAIN^{tetra} – A Conceptual Communication Model for Participatory Processes

Accountability and stakeholder consultation challenges the municipal authorities to develop transparent and efficient communication systems, which facilitate information sharing and feedback as well as procedures for participation, or representation, of different stakeholders in various decision-making processes (see Chapter 7, Section 1.7). Since such approaches are very different from prevailing technocratic program and implementation approaches, it will take time for good practice principles to emerge and be shared among various local authorities and their respective departments and agencies. Participative planning initiatives are both laborious and time-consuming, at least in the initial stages. It involves communication and learning to shape a common platform for committed further action of the actors and stakeholders involved. The challenges of communication will be discussed below in introducing a conceptual model developed at Chalmers School of Architecture (Kain 2003).

Planning is the preparation of intended interventions in a local situation, may it be on a more structural or detailed level. A local situation is constituted by the social activities taking place, framed by specific local conditions and influenced by external driving forces (legislation, market, etc) and other relations to the surrounding world at local, regional, and global levels. The specific local conditions are related to human made structures (e.g. buildings, construction and technical infrastructure), social institutions (e.g. formal and informal social organizations and power-relations), as well as the natural systems (e.g. flows and ecosystems dependent of geology and climate) of a place. These interrelated dimensions are labeled Artefact (A), Institution (I), and Nature (N), respectively, in the MAIN^{tetra} model of Figure 5-5. The bottom triangle of the model (AIN) can be regarded as representing the reality of a place. This is the field of investigation and analysis in the planning process to better understand the special dynamics of a local situation and the projected feasibility and impact of different interventions.

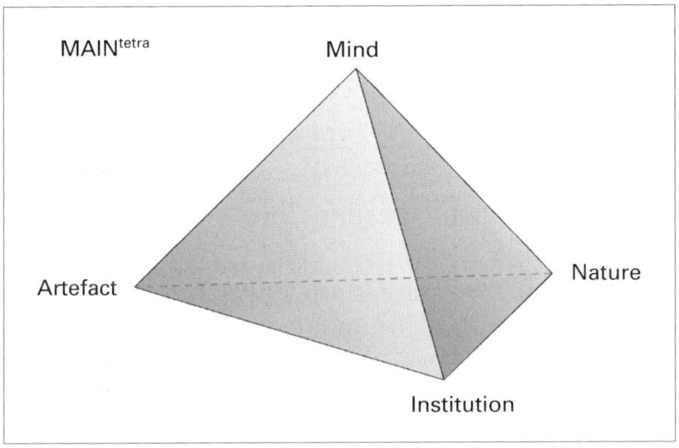

Figure 5-5. The MAIN^{tetra} conceptual model (Kain 2003)

It can be concluded that besides the need for a strong and equitable institutional framework, partnership and participation arrangements require methods and tools that can support the communication and learning of the different participants. This includes the proper management of uncertainties and power-relations always involved in planning and decision-making processes. This is the scope of an extensive debate in planning theory concerning the role and required skill of professional planners (see for instance the work of scholars like John Forester, Tore Sager, Patsy Healey, Bent Flyvbjerg, and Kimmo Lapintie during the last decade). Here, only three main challenges will be mentioned (further developed in Malbert 1998, Kain 2003):

- The need for conscious process design and process facilitation
- The need for relevant methods and tools for communication and decision-making support
- The need for working links and coordination between informal, and sometimes temporary, forums for knowledge development and formal arenas for decision-making and conflict management

5.3 Institutional Capacity – Regional and Global Partnerships and Networking for Sustainable Urban Development

As a demonstration of regional and global partnerships for sustainable urban development since the 1992 World Summit, the formulation of environment reports have been piloted in a variety of developing country cities based on the DPSIR framework (driving force, pressure, state, impacts and response) in the past decade. The program *Cities for Environmental Reporting On the Internet* (CEROI) was implemented as a collaboration between UNEP, ICLEI and various stakeholders within the respective cities, including the local municipal councils. In South Africa, Johannesburg, Pretoria, Cape Town and Durban were selected as pilot cities. Once completed, the reports are posted in the Internet to enhance access, exchange lessons and insights and compare the cities studied with each other. The spatial dimension of such monitoring and reporting is systematically captured in GIS format (see for example Greater Johannesburg Metropolitan Council 2000).

Today, new knowledge is not only created by specific research on new topics, but also by merging existing knowledge. Another task is to transform knowledge to action for a more sustainable world. Networking is an abstract form of collaboration and its organization may potentially be free of hierarchies. It is a voluntary collaboration among stakeholders, institutions, and authorities facing similar challenges for future development, all aiming at achieving sustainable living conditions.

Networking aims at putting together knowledge, creating synergies, and using resources more efficiently. Cities and regions are not islands and cannot act in isolation. Cities must look for collaboration with other communities worldwide, which have similar needs that can be met more easily, cheaply and successfully if jointly undertaken. Benchmarking and exchanging information with other cities can promote excellence in planning. In addition, best practices can be exchanged and learned from.

The Internet can serve as a tool for virtual and intercommunity cooperation aimed at achieving sustainable development for cities and regions. Some examples are the African Sustainable Cities Network, Eco-City Network, Smart Communities Network, The Cities Alliance, Citizens Network for Sustainable Development, The International Council for Local Environmental Initiatives (ICLEI), The International Network for Urban Development and Stockholm Partnerships for Sustainable Cities. These networks can work at national, continental, or global levels.

5.4 Corporatization, Privatization, and Public-Private Partnerships as Management Strategies for Infrastructure and Service Provision

The growth of cities and the related complexities in the governance structure demands constantly evolving strategies to manage the provision of infrastructure and services. In this context, a variety of alternative approaches have emerged especially in the last two decades (Wekwete 1997). Three key alternative strategies currently practiced even in developing country cities are corporatization, privatization and public-private partnerships. In all three strategies, a service level agreement constitutes the primary mechanism through which the performance of the entity is monitored and reviewed.

Corporatization entails the granting of operational autonomy to a service/infrastructure delivery entity, which is fully owned by the state or municipal authority. The budget of this entity would therefore constitute grants from the public sector as well as revenues generated from provided services. The primary aim of corporatization is to enhance operational efficiency and minimize bureaucratic control.

Privatization entails the provision of infrastructure or services, for example roads, water, electricity or waste removal, by a private organization under contract with the national or municipal government. One of the key mechanisms of this strategy is the concessionaire process, which outlines the scope of operation, tariff levels and geographic coverage of the contract as well as its duration.

In addition, *public-private partnerships* constitute a joint investment between the private sector and national or municipal government departments in service and related infrastructure.

One other critical component in sustainable urban development strategies, especially in developing countries, is the role of non-governmental and community-based organizations (NGOs and CBOs). With the additional resources they generate, internally or externally, and their investment in communities, they complement the role of the state in service delivery, especially for infrastructure and economic opportunities in community-based initiatives.

The City of Johannesburg demonstrates successful examples of corporatization. Since 2002, most of the infrastructure and services within the city are provided by corporatized entities. These include "Pikitup" (City of Johannesburg 2004), Johannesburg Water and City Power for waste, water and electricity services respectively. The Council has also established a variety of agencies mandated to pursue developmental initiatives in a pro-active approach (City of Johannesburg 2003b).

6. TOWARDS INTEGRATION OF STRATEGIES AND PROGRAMS

One of the greatest challenges in formulating and implementing sustainable urban development strategies in developing country cities relates to realizing an integrated approach, which can reconcile conflicting priorities as well as maximize the synergy between complementary strategies and programs (see Chapter 7, Section 1.6). For example, the delivery of new housing stock, which integrates energy and water efficiency, would be best achieved through an integrated approach between the housing and energy and water portfolios of the different spheres of government. The challenge arises from overcoming constraints in relation to costs as well as difficulties in interdepartmental partnerships in policy and program planning. The absence of good practice examples and the fear of losing departmental autonomy inhibit innovativeness in this regard. The major facets of this issue have been covered in detail in Section 5 above. This section explores the principles of LA 21 and integrated development planning as approaches towards integrating sustainable urban development strategies at formulation and implementation as well as monitoring stages.

6.1 Local Agenda 21 and HABITAT Agenda as broad Integration Frameworks for Sustainable Urban Development Strategies

The implementation of Chapter 7 of *Agenda 21* (UN 1992) and *Habitat Agenda* (UNCHS 1996) is supposed to be realized through the practice of *Local Agenda 21* (LA21) processes at the local authority level, which formulate and implement integrative plans for sustainable development. Therefore, LA21 should reinforce the other two documents as well as any other which advocates for sustainable development at a local scale.

In the evolutionary process of urban planning management and development, integrated development planning has emerged as a contemporary approach. This entails a transition from sector planning in isolation (with land use planning and spatial master plans as the key components) to a comprehensive and synthesized planning where all key sectors, including finance/budgeting and the environment, are synergistically combined in a single urban plan anchored on a common vision of the city or settlement.

The comprehensiveness of the integrated plan would thus depend on the comprehensiveness of the visioning/backcasting process. For example, if sustainability issues are addressed and prioritized during the vision derivation process, the integrated plan is unlikely to aid strategies and mechanisms that systematically enhance sustainability levels.

Even though LA21 and integrated planning emphasize the consultation process for envisioning and plan formulation and implementation, integrated planning may not necessarily promote sustainability comprehensively if key issues are ignored at the initial stages. Taking the two frameworks and conventional planning practice, one can identify five key approaches to planning in urban settlements in developing countries:

- Unmanaged or unplanned urban growth (common in countries subject to civil unrest or natural calamities such as drought).
- Conventional planning (sector plans done in isolation with spatial planning as the key component, most of which are local interpretations of a national development plan). This is common in most municipalities of developing countries.
- Conventional planning with isolated LA21 initiatives and projects. The presence of LA21 initiatives reflects some form of sustainability awareness and interest but possibly the absence of mainstream access. This is evident in several cities in developing countries especially where cooperation with external stakeholders is involved.
- Partial sustainable urban development approach (integrated planning with some input from LA21 initiatives and projects). This is only evident in extremely few cities in developing countries and most often entails strong collaboration with external stakeholders from UN or developed countries.
- Comprehensive sustainable urban development approach (LA21 principles as the overall framework for integrated planning: Merges the elements of Chapter 7 and 28 of *Agenda 21* as well as those of *Habitat Agenda* into the planning and implementation/management process; Keiner 2003). The authors are not aware of any city in developing countries that is currently practicing sustainable urban development at this level.

Figure 5-6 represents the last ideal approach for integrating strategies for sustainable urban development.

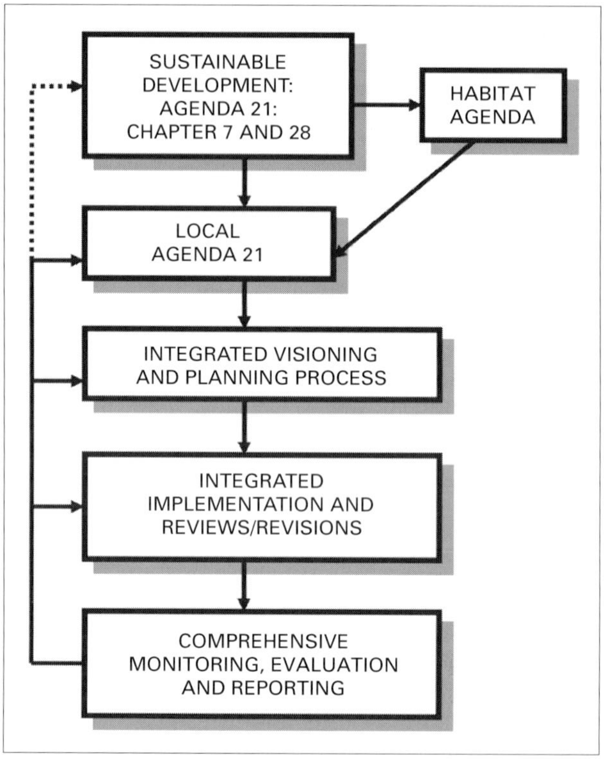

Figure 5-6. Comprehensive approach for integrating sustainable urban development strategies.

6.2 Integrated Development Planning

Following the transition to democracy in 1994, South Africa was confronted with the tremendous challenge of transforming the apartheid system of governance to a more democratic system capable of confronting 21st century challenges in governance and development. The ANC-led government captured the framework for confronting the challenge in two key documents: the *Reconstruction and Development Program* (RDP; ANC 1994) and the *Constitution of South Africa* (RSA 1996).

The system of governance was legislated through the Constitution. In particular, the setting of national, provincial and local authorities as separate but cooperating spheres of government provided for the relative autonomy of local authorities. This was later elaborated through the Municipal Structures Act of 1998, which elaborated on the mandate, and operational procedures of municipal governments.

There was also a critical need for a systematic approach to planning and urban development in order to undo the spatial and socioeconomic legacies of apartheid. Early attempts after 1994 started with the ANC's intention of facilitating expedited housing delivery to tackle the severe backlog. Unfortunately, such attempts failed to deliver significant impacts and the resultant housing ended up being located on the periphery of key urban nodes just as in the apartheid era.

The promulgation of the Municipal Systems Act of 2000 was aimed at supporting efficient and systematic development planning and implementation. It was this piece of legislation under the Department of Provincial and Local Government (DPLG), which gave Integrated Development Planning (IDP) authority as the common framework for planning at provincial and local government levels.

The key requirements for the formulation and implementation of IDP are as follows:

- Integration of all sector plans and planning processes into one plan addressing the economic, spatial, social development as well as environmental management
- A municipal budget which comprehensively addresses the administration of the municipality as well as funding for proposed programs and projects
- Stakeholder/community participation during formulation and implementation processes
- Integration of performance issues and criteria by which implementation and related outcomes will be monitored and evaluated
- Annual review and revisions
- Appraisal and comments by provincial departments responsible for IDP and municipal operation

The IDP process developed action plans for the eleven regions of the city. Despite poor attendance at IDP meetings, the overall priorities that were determined nevertheless cover broad concerns that almost all communities can identify with. The six strategic priorities of IDP are listed in Table 5-3.

Table 5-3. Interrelationship between IDP strategic outcomes and Mayoral Priorities for Johannesburg

Strategic Outcomes of IDP	Mayoral Priorities
– Affordable, sustainable customer-focused delivery	– Service delivery excellence
– Economic growth and development	– Economic development and job creation
– Safe and secure city	– Bylaw enforcement and crime prevention
– High quality built and natural environment	– Inner city renewal
– Citizens enjoy an improved quality of life	– HIV/AIDS prevention
– Financial viability and strength, organizational development and excellence	– Good governance, customer care and *Batho Pele* "People First"

This exercise culminated the formal adoption of IDP by the city at the beginning of 2002. In addition, *Key Performance Areas* (KPAs) were determined and these are to be monitored using *Key Performance Indicators* (KPIs). To-date, the city has completed the annual review of the IDP as well as the formulation of the second cycle of IDP and SDF for 2003/04.

Although urban development planning has been plagued with a plethora of disjointed pieces of legislation, the provisions of the Constitution, Municipal Structures Act and Municipal Systems Act have gone a long way towards democratizing municipal governance for development in South Africa. Even though the first two cycles of the IDP have not been as integrated and consultative as required, it is anticipated that future cycles will incrementally improve on the foundation they have laid. An anticipated land use bill will further enhance spatial planning aspects of the IDP, as it will streamline the diverse and disjointed pieces of spatial planning legislation.

One of the key shortcomings of the IDP as it has been legislated is that it does not specifically require the application of sustainable urban development principles in the planning framework (see Chapter 7, Section 1.6). Similarly, although there have been attempts to draw comparison between IDP processes and outcomes with *Agenda 21* principles (Coetzee 2002), no local authority has adopted *Agenda 21* as the framework for its IDP, even in municipalities where *Local Agenda 21* initiatives and projects are ongoing.

Even though there has been no explicit attempt to carry out the practice of IDP in Johannesburg within the sustainable urban development framework, there exists a strong foundation for this to happen in the future. There are three key pillars for the foundation of IDP:

- Governance: as required under the Constitution and the Municipal Structures Act
- Planning: as required under the Municipal Systems Act with IDP and community consultation as key requirements
- Backcasting: as applied in the formulation of *Joburg 2030*

Integrated development planning might also be an excellent tool for curbing Santiago's major problems. In this regard, integrated planning could be applied as a process through which the city can establish a plan for the short, medium and long-term sustainable development of the city region and manage the environment in the process. The IDP process can be outlined as follows:

- Assess an area's current situation, including available resources, skills and capacities
- Assess the needs of the community and industry
- Prioritize these needs in order of urgency and importance
- Set goals to meet these needs
- Devise strategies to achieve the goals within a set timetable
- Develop and implement projects and programs to achieve key objectives
- Set targets so that performance can be measured
- Budget effectively with limited resources
- Regularly monitor and reassess the development program
- Make changes where necessary

Ultimately, the key to integrated development planning is that it considers not just one sector (e.g. transport) or one group of issues (e.g. environment), but instead brings together important sectors, issues and concerns into an "integrated" whole. IDP also requires the national, regional and local government, industry as well as individual residents, non-governmental organizations and civic groups, the private sector and other stakeholders to set goals and work together in an organized way to achieve them. Such integration requires holistic thinking. Issues and sectors are looked at in relationship to each other, not in isolation. This enables the very best use of resources to achieve development aims (Marjanovic and Cavric 2003).

7. MONITORING, EVALUATION AND REPORTING ON SUSTAINABLE URBAN DEVELOPMENT STRATEGIES AND PROGRAMS

Central and local governments must be aware that urban development planning and physical planning is an ongoing process facilitating the provision of a sound social, economic, and ecological environment for future generations. Monitoring and controlling concepts (see Chapter 7, Section 1.8) help to guide and shape the urban development and planning processes.

Monitoring and evaluating sustainable urban development strategies and programs is crucial as it facilitates the tracking of outcomes in order to inform necessary revisions. The prioritization of sustainable urban development issues to be monitored and tracked in relation to the impacts of various strategies and programs can only materialize from a consultative process. This is strongly linked to the issue of indicators as discussed in Chapter 6. Another key challenge under monitoring and reporting relates to the quality of data and their analysis, especially in view of facilitating comparisons over time and among cities.

Between 2000 and 2002, Johannesburg underwent a major institutional and planning transformation in response to changes in national legislation as well as internal pressures for institutional reform. In order to facilitate the change, the city implemented a major data collection and analysis process towards helping the city in benchmarking itself and establishing key gaps between its current state and its vision for the future. Data collection and reporting was done under three major categories (City of Johannesburg and Monitor Company 2000):

- Economic development (manufacturing, financial services, trade and logistics)
- Infrastructure and services (housing, electricity, water and sanitation, solid waste, roads and storm water, environment and property)
- Human development (demographics, safety and security, education, skills development, health, HIV/AIDS, social and community services)

This period produced the most comprehensive mapping of the city in terms of data categories and intensity as well as geographic coverage and regional disaggregation. The outputs of the data capture and analysis as well as the strategies for addressing the major gaps identified were consolidated into the *iGoli 2010* report. The report, together with national legislation such as the Municipal Structures Act of 2000, was used as the framework for institutional restructuring in 2002. However, the implementation of the recommended strategies could not be formally adopted because a ten-year period was deemed unrealistic. Instead, a more feasible thirty-year timeframe was adopted.

The process of transforming the *iGoli 2010* report to the thirty-year framework generated *Joburg 2030* as the official framework. It is this report, which captured Johannesburg's vision of being an *African World Class City*. However, *Joburg 2030* took a strong bias towards economic growth and global competitiveness in contrast to the broader range of local and grassroots issues which *iGoli 2010* had prioritized (see City of Johannesburg and Monitor Company 2000 and City of Johannesburg 2002). *Joburg 2030* targets at putting in place strategies that will ensure an improved labor force with appropriate skills, as measured by increased numeracy and technological literacy. *Joburg 2030* concedes that national policy processes largely direct the local government, making it primarily a "policy taker," rather than a "policy maker." This has prompted the city to investigate how it can create supportive environments as an agent for economic growth. It is envisaged that increased Gross Geographic Product (GGP) will result in better standards of living and an improved quality of life, as measured by the *Human Development Index*. Furthermore, the city wants to ensure that increased GGP is shared by all; increases in salary will therefore be accompanied by increased parity. Increased GGP will also mean increased revenue that will enable the city to provide better facilities. Other research conducted for *Joburg 2030* identified high levels of crime and inappropriate skills as the major impediments in investment decisions. *Joburg 2030* has adopted these concerns as "five star priorities," in order to gain increased business confidence and increased investment. For example, due to increased disposable incomes, there is expected to be an increased demand for open spaces, and rivers and dams will be restored for outdoor relaxation. *Joburg 2030* also aspires to suburbanize black townships along the lines of white suburbs. It is believed that this will lead to the development of a property market in traditionally black townships. Poorer communities will be concentrated in special needs areas. The apartheid city was a divided city, and these goals raise questions as to what extent the two parts are to be integrated. At the very least, racial segregation must not be replaced by separation along economic lines.

Subsequent planning through the *Integrated Development Planning* process has benefited significantly from the two-year benchmarking period, especially with regard to the rich pool of data and analysis that was produced. The *iGoli 2010* process also initiated the broad stakeholder consultation review, which was enhanced through subsequent IDP cycles. Due to the new Constitution and governance system of democratic South Africa, IDP has become a legislated requirement of each local authority. The strong commitment of governments at different spheres (local, provincial and national) to ensure that IDPs are generated and implemented effectively will definitely ensure the nurturing and maturing of a systematic planning practice in South Africa and Johannesburg in particular. However, in spite of the strong alignment between IDPs and *Local Agenda 21* principles and processes, there has not been an official recognition of the opportunity to apply sustainable urban development as the overall framework for the IDP process and its outcomes. Consequently, urban sustainability strategies, including tools and mechanisms, have not been explicitly applied in the conceptualization of the IDPs (see Chapter 7, Section 1.6).

The other major advantage of the *iGoli 2010* process is that it established a framework for monitoring and reporting progress along the key strategic thrusts identified for action. However, this opportunity seems to have been missed due to the extremely narrow focus of the key performance areas and indicators defined in subsequent IDP cycles. The IDP indicators tend to bias the monitoring and reporting efforts towards the operational and management issues under the portfolios of various political and executive leaders of the city. The different predilections include the following:

- Customer perspective (customer-focused service delivery, safety and security, improved quality of life for citizens)
- Financial perspective (customer-focused service delivery, financial viability and strength)
- Internal business process perspective (safety and security, quality built environment, financial viability and organizational development)

Consequently, the opportunity of systematic tracking and reporting on the broader range of indicators as envisaged under *iGoli 2010* has been significantly undermined. Similarly, the opportunity of using such indicators in informing future cycles of Johannesburg's IDP towards sustainable urban development may have been missed.

The *State of the Environment Report* (City of Johannesburg 2003c), which is one of the major inputs to the environment sector plan, comes closest to identifying the broader sustainable urban development challenge and opportunities. The document actually sets a strong base for the status quo and future monitoring of key sustainability indicators (environmental, socio-economic, cultural and institutional). The document applies the DPSIR model (driving force, pressure, state, impacts and responses) of data capture and reporting. It is clear that with the regular gathering and reporting of such data, one can easily track sustainability trends of the city. However, there are no systematic mechanisms or strategy to link the monitoring outcomes to the other sector plans to ensure action aimed at improving the indicators themselves in an integrated and synergetic approach. In other words, there has not been an explicit effort towards closing the loop of sustainable urban development – planning, implementation, monitoring/evaluation and back to planning.

One can therefore conclude that from a legislative and institutional point of view, Johannesburg has a strong base for embarking on sustainable urban development. Recasting *Joburg 2030* and future cycles of IDPs (including related SDFs and other sector plans) with the broader framework of urban sustainability would be a primary step in the right direction. This could be followed by expanding the key performance areas and indicators of the IDP (KPAs and KPIs) of the city's score card to include both operational and management areas or indicators as well as broader urban-sustainability indicators to facilitate and support the monitoring and evaluation for both categories of issues rather than just the management ones. Johannesburg must refocus its attention on this problem in order to sustain its global and regional competitiveness and also ensure that grassroots needs will be met in a sustainable manner over time.

8. CONSOLIDATION AND CONCLUSIONS

As the key frameworks underpinning the sustainable development vision and paradigm, *Agenda 21* (Chapter 7 and 28 in particular) and *Habitat Agenda* can be viewed as the outcome of a global backcasting process achieved through UN-brokered scientific and political consultations and negotiations. However, the expectation that these global frameworks would then serve as references for national and local strategies for sustainable development has not been systematically realized, especially in developing countries and their respective cities.

For example, national and local backcasting processes in Botswana and Johannesburg (*Vision 2016* and *Joburg 2030* respectively) have been formulated without explicit reference to the sustainable development frameworks. Equally, national and local planning in most countries continues to be based on sectoral plans with minimal provisions for vertical and horizontal integration. The bias of such visions and plans with socioeconomic challenges as the top priority and local environmental concerns as a secondary priority has marginalized issues of intergenerational equity and global environmental concerns such as global warming, overproduction and overconsumption.

The case study cities provide significant manifestations of this phenomenon. In Botswana for example, even though a vertical integration of planning is emerging, there is a strong bias towards spatial and economic planning, policies and strategies, with limited attention towards resource and environmental implications of development for future generations. The horizontal integration of sectoral plans is also extremely weak both at national and local levels. For example, the expectation that environmental and resource conservation should provide the overarching framework for economic, spatial and social development planning, and not treated as an isolated sector or program, does not seem to have been recognized or accepted in policy and planning practice in Botswana and Gaborone in particular.

In Santiago, attempts to link spatial and environmental planning with local impacts of air pollution have been undermined by the absence of effective legislative and institutional frameworks. Whereas Johannesburg has taken advantage of the integrative approach required by various pieces of legislation in South Africa, it has fallen short of recognizing sustainable development as the anchor framework for its *Joburg 2030* and its *Integrated Development Plan*. In particular, the vision and the IDP downplay environmental and resource management issues in favor of economic growth and conventional competitiveness.

In all three case study cities, there is no evidence of sustainability-linked monitoring, evaluation and reporting. Instead, monitoring and reporting for a few selected items seems to be the norm. In Santiago, air quality monitoring and reporting has been prioritized due to the local impacts especially regarding threat to health. In Johannesburg, the IDP process has prioritized monitoring and reporting of operational and management performance according to lines of responsibility rather than broader sustainability goals, for which no one within the Council is directly responsible for. In Botswana, concern for vulnerability due to a mining sector dependency, especially the diamond industry, has led to the prioritization of monitoring and reporting along this issue.

It is therefore unlikely that benchmarking and progress towards sustainable urban development will be systematically pursued or realized in the case study cities in the next years. However, all three cities exhibit several pieces of policies, legislation and plans at national and local levels on which a comprehensive approach to sustainable urban development could be based. Political will and institutional capacity at national and local levels are therefore likely to remain the key determinants of progress towards sustainable urban development in the case study cities as well as most cities in developing countries. In particular, the ability among key stakeholders to visualize and prioritize the solution of socioeconomic problems within a resource and environment conservation framework will be a vital political milestone in this direction.

REFERENCES

African National Congress (ANC) 1994. *The Reconstruction and Development Programme*. Johannesburg: Umanyano Publications.

Azar, C., J. Holmberg, and S. Karlsson 2002. *Decoupling – Past Trends and Prospects for the Future.* A report for the Swedish Environmental Advisory Council 2002:2. Gothenburg: Chalmers University of Technology.

Coetzee, M. 2002. *Local Pathway to Sustainable Development in South Africa.* Pretoria, South Africa: CSIR.

City of Johannesburg 2002. *Joburg 2030 – Full report.* http://www.joburg.org.za/joburg_2030/2030_documents.stm, retrieved on March 2, 2004.

City of Johannesburg 2003a. *Spatial Development Framework 2003/04*. Johannesburg.

City of Johannesburg 2003b. *Integrated Development Plan, 2003/04*. Johannesburg.

City of Johannesburg 2003c. *State of Environment Report 2003*. Johannesburg.

City of Johannesburg 2004. *Pikitup*. http://www.pikitup.co.za/, retrieved on March 30, 2004.

City of Johannesburg and Monitor Company 2000. *iGoli 2010: Final Integrated Human Development Story and Final Integrated Story for the Mayoral Committee*. Johannesburg.

Du Plessis, C., D.K. Irurah, and R.J. Scholes 2003. The Built Environment and Climate Change in South Africa. *Building Research and Information* Vol. 31, No. 3–4: 240–256.

Fudge, C. and J. Rowe 2000. *Implementing Sustainable Futures in Sweden*. Stockholm: The Swedish Council for Building Research.

Government of Botswana 1998. *Botswana National Settlement Policy*. Government Paper No. 2 of 1998. Gaborone.

Grameen Foundation USA 2003. *Report Bangladesh 2000–2002*. Washington DC.

Greater Johannesburg Metropolitan Council 2000. *State of the Environment*. http://ceroi.net/reports/johannesburg/csoe/Default.htm, retrieved on March 3, 2004.

Higinbotham Holcombe, S. 1995. *Managing to Empower: The Grameen Bank's Experience of Poverty Alleviation*. London: Zed Books.

Irurah, D.K. (Ed.) 2000. *Environmentally Sound Energy Efficient Low-Cost Housing for Healthier, Brighter and Wealthier Households, Municipalities and Nation*. Final Report. Technical Report to the Environmentally Sound Low-Cost Housing Task Team and USAID. Pretoria.

Kain, J.-H. 2003. *Sociotechnical Knowledge: An Operationalised Approach to Localised Infrastructure Planning and Sustainable Urban Development*. Dissertation, Department of Built Environment & Sustainable Development. Gothenburg: Chalmers University of Technology.

Keiner, M. 2003. Neue Formen der Planungsbeteiligung als Folge des Lokalen Agenda 21-Prozesses. *ETH e-Collection*. http://e-collection.ethbib.ethz.ch/cgi-bin/show.pl?type=bericht&nr=306, retrieved on May 5, 2004.

Malbert, B. 1998. *Urban Planning Participation: Linking Practice and Theory*. Gothenburg: Chalmers Architecture.

Marjanovic, P. and B. Cavric 2003. *Environmental Management Systems and Transition to a Sustainable Development: The Role of Integrated Planning and Adoptive Management Strategies in Economies in Transition*. 1st International Symposium on Environmental Management. Zagreb, Croatia, 1–4 October 2003, Track 1. Environmental Management Trends and Results.

Meadows, D. L., J. Randers, and W.W. Behrens III. 1972. *The Limits to Growth: A Report for the Club of Rome's Project on the Predicament of Mankind*. New York: Potomac Associates.

Ministerío de Vivíenda y Urbanismo (MINVU) 1994. *Plan Regulador Metropolitano de Santiago (PRMS)*. Santiago.

Presidential Task Group for a long-term Vision for Botswana 1997. *Vision 2016 – Towards Prosperity for All*. Gaborone: Government Printer.

Republic of South Africa 2003. *National Environmental Management: Air Quality Bill*. http://www.gov.za/gazette/bills/2003/25289a.pdf, retrieved on March 30, 2004.

Republic of South Africa 2000a. *Local Government Municipal Structures Act, 2000*. Act No. 33 f 2000 (An amendment to the *Local Government Municipal Structures Act 1998*. Act No. 117 Of 1998). http://www.gov.za/acts/1998/a117-98.pdf, retrieved on March 30, 2004.

Republic of South Africa 2000b. *Municipal Systems Act 2000*. Act No. 32 of 2000. http://www.gov.za/acts/1998/, retrieved on March 30, 2004.

Republic of South Africa 1996. *The Constitution of the Republic of South Africa, 1996*. Act No. 108 of 1996. http://www.gov.za/acts/1996/, retrieved on March 30, 2004.

Sassen, S. 2002. Locating Cities on Global Circuits. *Environment and Urbanization*. Volume 14, No. 1: 13–30.

Schmid, W.A. et al 1996. *Raumplanung Grundzüge, Eine Einführung für Ingenieure*. Skript zur Vorlesung von Prof. Dr. W.A. Schmid. Institut für Orts-, Regional- und Landesplanung, ETH Zürich.

Sustainable Cities 2002. Sustainable Cities: Globalisation and Cities. *Environment and Urbanization*. Vol. 14, No. 1: 3–12.

Sustainable Cities 2000. Towards More Pro-Poor Local Governments in Urban Areas. *Environment and Urbanization*. Vol. 12, No. 1: 3–11.

Sustainable Cities 1992. Meeting Needs, Reducing Resource Use and Recycling, Re-Use and Reclamation. *Environment and Urbanization*. Vol. 4, No. 2: 3–8.

Suzenet, G., A. Tal, and D. Boymanns 2002. Sustainable Water Management for the City: Technologies for Improving Domestic Water Supply. *Built Environment* 28/2: 138–163.

Thalmann, B. 2002. *Analytical Tools and Models for Sustainable Urban Development – A Framework for Spatial Monitoring and Controlling in Gaborone (Botswana)*. Diploma Thesis, diploma course Rural Engineering, ETH Zurich.

UNCHS (Habitat) 1996. *Habitat Agenda And Istanbul Declaration:* Second UN Conference On Human Settlements. Istanbul, Turkey 3–14 June 1996. http://www.unchs.org/unchs/english/hagenda/ist-dec.htm, retrieved on May 9, 2004.

UNEP-CBD (United Nations Environmental Program) 2003. *Convention on Biological Diversity. Convention Text.* http://www.biodiv.org/convention/articles.asp, retrieved on June 23, 2004.

UNFCC (United Nations Framework Convention on Climate Change) 1992. *Kyoto Protocol to the United Nations Framework Convention on Climate Change*. http://unfcc.int/resource/convkp.html, retrieved on March 30, 2004.

United Nations (UN) 1992. *Agenda 21: The UN Program of Action from Rio*. http://www.un.org/esa/sustdev/documents/agenda21/english/agenda21toc.htm, retrieved on March 30, 2004.

United Nations (UN) 2000. *Millennium Development Goals: Millennium Declaration.* Resolution adopted by the General Assembly. http://www.un.org/millenniumgoals/, retrieved on May 7, 2004.

United Nations (UN) 2002. *Johannesburg Plan of Implementation: A Report of the World Summit on Sustainable Development*. Johannesburg, August 26 to September 4, 2002. http://www.johannesburgsummit.org/html/documents/summit_docs/131302_wssd_report_reissued.pdf, retrieved on March 30, 2004.

Wang, C.K. and P. Guild 1995. The Strategic Use of Organizational Competencies and Backcasting in Competitive Analysis. *ISBM Report* No. 7-1995. http://www.ebusiness.xerox.com/isbm/dscgi/ds.py/Get/File-66/24-1995.pdf, retrieved on March 30, 2004.

WCED (The World Commission on Environment and Development) 1987. *Our Common Future.* Oxford: Oxford University Press.

Weizsäcker, E.U., H. Lovins, and A. Lovins 1997. *Factor Four: Doubling Wealth, Halving Resource Use*. London: Earthscan.

Wekwete, K.H. 1997. Urban Management: The Recent Experience. In C. Rakodi (Ed.) *The Urban Challenge in Africa – Growth and Management of its Large Cities.* Tokyo, New York, Paris: UNU Press: 527–552.

Yunas, M. 1999. The Grameen Bank. *Scientific American*, 281, No. 5: 114–119.

Chapter 6

INDICATORS FOR SUSTAINABLE URBAN DEVELOPMENT

P. Christopher Zegras[1], Iván Poduje[2], Whitney Foutz[1], Eran Ben-Joseph[1], Oscar Figueroa[2]
[1]*Massachusetts Institute of Technology*, [2]*Pontificia Universidad Católica de Chile*

Abstract: This chapter explores the critical issue of measuring sustainable urban development (SUD) via the use of indicators. The chapter begins by situating indicators within the broader urban planning process, showing how they help measure goals and objectives, define evaluation criteria and monitor progress. Recognizing the inherent complexity of the sustainability concept, the chapter then presents the hierarchy of information generally used, in the measurement process. Several different approaches and techniques for measuring sustainability – via indicators and indices – are presented, and building on these precedents, the sustainable indicator prism is proposed as a means for structurally guiding and understanding the informational and geographical scales implicit in the multidimensionality of sustainable urban development (SUD). The chapter recognizes that a "one-size-fits-all" approach to SUD indicators is not possible; key challenges include the lack of responsible authority and the lack of place-specific data. Ultimately, indicators must reflect political reality, information availability and a relevant scale of analysis.

Key words: Indicators, indices, performance measures, sustainability indicators

1. INTRODUCTION

1.1 The Role of Indicators in the Urban Planning Process

In a serious attempt to make the concept of sustainable urban development operational, we not only face the challenges of status quo

M. Keiner et al. (eds.), From Understanding to Action, 157–189.

pressures, powerful urbanization trends, difficulties in implementing the appropriate policies and programs, etc. Another challenge relates to possibly more fundamental questions, particularly, "How do we know how sustainable our urban development is?" and, perhaps even more importantly, "How do we know that our proposed interventions will lead us towards sustainable urban development?" Answering such questions requires efforts to quantify the various dimensions of sustainable urban development and predict how they will change over time. Indicators (or performance measures) aim to assist us in this regard.

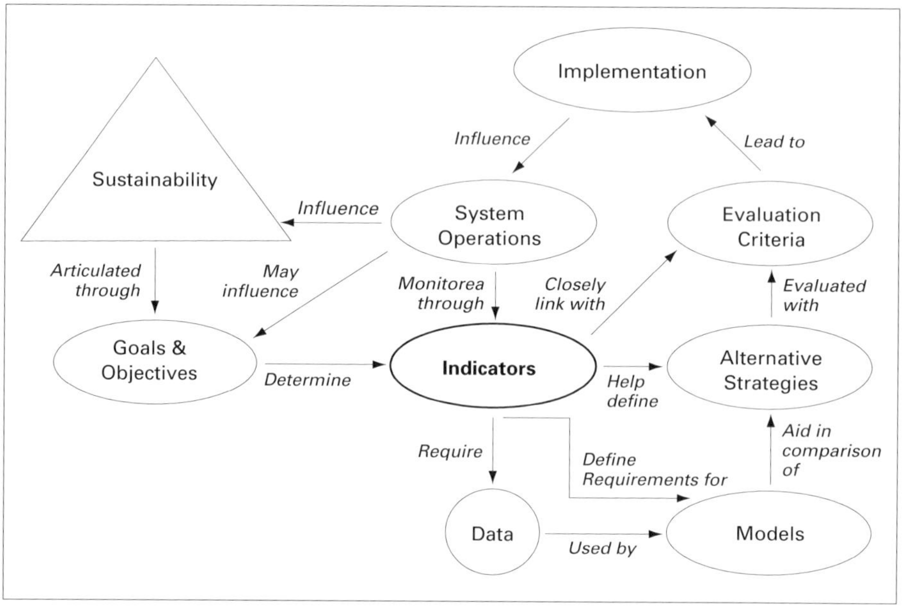

Figure 6-1. Indicators in the Urban Planning Process (Source: adapted from Meyer and Miller 2001)

The schematic in Figure 6-1 shows a simplification of how indicators relate to other key aspects of the planning process. As can be seen, indicators, which rely on some form of data (see Sections 2 and 3.8 of this chapter), reflect the overall goals and objectives of sustainable urban development (SUD), help define alternative strategies for SUD, are closely linked to – though not necessarily identical to – evaluation criteria and are also used to help monitor and control system operations (see Chapter 7, Section 1.8 and Keiner 2002). Indicators also play an important role in defining the requirements for the models used to evaluate different strategies or interventions. The relations depicted in Figure 6-1 do not include every aspect involved. Public participation, for example, plays a role in many of the actions shown but is not referred to explicitly. Furthermore, it remains questionable whether such a logical process is ultimately followed in any actual planning activity, since politics, institutional inertia, and other practical considerations inevitably and necessarily influence the ultimate approach.

This Chapter focuses primarily on indicators and, due to their underlying importance, data. We need to point out at the outset that indicators represent *abstractions* of reality: An indicator provides an approximation of a condition that we are concerned with. The interest lies, however, not on the indicator itself (e.g. the ratio of sum of all floor spaces to a plot area) but on the inferred conclusions (e.g. settlement compression). It is necessary that the relationship and dependency between the indicator and the derived conclusions are clear (e.g. the higher a ratio of floor spaces to a plot area, the more compressed the settlement is). The usefulness of indicators depends thus on their suitability, appropriateness and the precision of the relation to the conclusions. The selection of indicators follows comprehensible criteria, but due to the varying interpretations of their qualitative results, indicator based analysis remains subjective to a certain degree (Keiner 2002). In other words, indicators give us information that can, ideally, help us understand how sustainable a system is, what the pressures are, etc. Indicators used in assessing "sustainable urban development" depend on multiple factors, but most particularly the *scale, scope*, and *time frame* of the analysis:

- Scale (e.g. metropolitan, intrametropolitan, neighborhood, site-specific)
- Scope (e.g. air quality, energy consumption, transportation, water consumption)
- Time frame (e.g. short, medium, long-term)

In general, the sustainability concept should cover all scales and scopes over the long-term. Sustainable urban development must apply at both the metropolitan and site scales and include all sectors (e.g. sanitation, transportation, housing, etc.). Furthermore, the very word implies a long-term view (i.e. "Can our social, economic and environmental development be 'sustained' in time for present and *future* generations?"). In practical terms, however, we cannot consider all impacts, at all scales, for eternity – it is clear that since planning involves the projection of and abstraction of future conditions, it must be flexible enough to accommodate differences in what actually takes place.

1.2 System Complexity and Boundaries

The multiple scales, scopes, and time frames of urban systems analysis underlies the overall complexity of the sustainability concept; we are looking at systems that are – as Innes and Booher (1999) note –

"[...] adaptive and self-organizing, with [their] components free to co-evolve in response to changes in each other and, as a whole, changing in response to external conditions."

Sussman (1999) uses the term CLIOS (Complex, Large-scale, Integrated, Open Systems) to refer to such systems, which are characterized by the following: interrelated units, for which we have imperfect knowledge about the relationships involved; large, pervasive and enduring impacts; integrated subsystems coupled through feedback loops; and openness in terms of all-encompassing, including social, political and economic aspects. Such systems are not only difficult to measure (i.e. via indicators) but also quite difficult to model.

At the same time, measurement implies some boundary requirement; indeed, the definition of sustainable *urban* development by itself introduces a boundary condition – the urban area. Such a boundary poses challenges since a city functions based on its interactions beyond its region (e.g. via trade, tourism, etc.) and produces impacts well beyond its borders. This is related to a city's role as an "enabler" of activity. Formally, cities owe their formation to the so-called "economies of agglomeration" (for more detail, see e.g. Glaeser 1998). The possibility may well exist that cities' wealth and consumer-orientation might lead to broader behavior that is "unsustainable" on a larger scale or over a longer time frame. Can metropolitan areas (and their own, bounded, sustainability) enable a larger, unsustainable *global* system? Ryan and Throgmorton (2003) offer a related observation in their comparison of sustainable transportation and land development in a German and a Californian urban case. They conclude that both cities are pursuing sustainable patterns of development in the local sense, or that their citizens and officials are aiming to sustain the features of their cities "which they admire." In the California case, however, the authors suggest that a community outside Chula Vista has opted to "sustain certain transportation and land use qualities which may not be sustainable in a global sense." This reflects the somewhat dichotomous situation between developing and developed world cities: In developing world cities, possibly the greatest sustainability threats comes from the living conditions within the cities (e.g., pollution, health risks, etc.), while for developed world cities (exemplified by places like Chula Vista), the greatest sustainability threats may well come from their global impacts (see e.g. Rees 1997).

Finally, in considering boundaries, we cannot ignore the somewhat arbitrary political units that typically define urban areas. Many cities today, like Gaborone, Santiago de Chile and Johannesburg, are multi-jurisdictional amalgams, with the urbanized area constantly expanding and encompassing new areas.

Leaf (2004) points out:

"Considering the possible disjuncture between the practice of planning and the processes and patterns of urbanization, especially rapid urbanization under conditions of limited societal resources, we need to ask whether the boundaries which define and delimit planning practice adequately fit with the overall parameters of urbanization, as understood spatially, temporally, and with regard to the institutional requirements of local governance."

Where does the "city" end? This is a critical issue not only for data collection and indicator development but – on a practical level – for many metropolitan management and coordination functions (e.g. land use, transportation, air quality, water quality, affordable housing provision; see Wheeler 2000). For example, Portney (2003), in a recent attempt to assess the degree to which U.S. cities are "taking sustainability seriously," considers cities to be a relevant unit of analysis both because they have to confront, head-on, many sustainability issues, such as waste disposal, and because city governments are close to the populace. At the same time, however, he notes that the city may be too small of a unit to make much of a difference for many of the larger problems. A related point is that ecosystems and governmental units rarely match up (see e.g. Gustavson et al. 1999), except in specific cases such as watershed or air quality management districts.

2. INDICATORS AND THE INFORMATION PYRAMID

Measuring any concept, such as sustainable urban development, first requires an operational definition. This definition should provide specific guidance on how the concept will be measured (Meier and Brudney 2002). For example, we can establish an operational definition for meeting air quality standards for fine particulate matter ($PM_{2.5}$) as follows: "Areas will be in compliance with the annual $PM_{2.5}$ standard when the 3-year average of the annual arithmetic mean $PM_{2.5}$ concentrations is less than or equal to 15 $\mu g/m^3$." In this case, the operational definition establishes how the concept, air quality compliance for fine particulates, will be measured: through mean $PM_{2.5}$ concentrations. In other words, we would use mean $PM_{2.5}$ concentrations as an *indicator* of compliance with air quality standards.

2.1 Indicators

Indicators serve a variety of purposes. They can help us better understand the current state of a system and its evolution, highlight areas for policy focus and indicate temporal improvements in conditions, thereby improving performance assessment by policymakers and the general public and ultimately enhancing management and accountability. Indicators can also serve to compare different systems (e.g. one city versus another city). Many metrics can be considered as indicators. For example, traditional economic measures, such as gross domestic product (GDP), inflation rates or unemployment rates, offer certain indications of the performance of an economic system. Social indicators, such as the literacy rate or the crime rate, help shed light on a society's social conditions. Generally, indicators can help "localize, measure and monitor conditions and change over time," but they "do not explain underlying causes" (OECD 1997).

When it comes to identifying and using indicators, there are two critical aspects to consider: *validity* and *reliability*. Validity refers to the accuracy with which the indicator measures the concept of interest. This relates, first and foremost, to the need to define indicators based upon the concept of interest and not vice versa. In other words, in using indicators we run the risk of naturally limiting our assessments to what is measurable, with the things we can measure ultimately forming the focus our efforts, irrespective of their ultimate relevance or importance.

There are four basic ways by which one can gauge the validity of an indicator (see Table 6-1). Even if an indicator meets these criteria, there is no "guarantee" of its ultimate validity. As Meier and Brudney (2002) point out, "Developing valid indicators of concepts is still an art;" indicator validity ultimately remains a subjective determination. The reliability of an indicator, on the other hand, tells us whether repeated measures of an unchanged phenomenon using the same indicator will give us the same value (i.e. it is not affected by anything other than actual changes in what is measured). Two principal threats to reliability exist: subjectivity, whereby the analyst's judgment influences the indicator value (e.g., the quality of a neighborhood); and imprecision, which refers to errors resulting from various factors such as sample size or imprecise definitions (Meier and Brudney 2002).

Table 6-1. Validity of indicators (Source: Meier and Brudney 2002)

Type of validity	Meaning
Face validity	Accepted by analyst as valid indicator of concept
Consensual validity	Accepted by numerous persons as valid indicator of concept
Correlational validity	Correlated strongly with other valid indicators
Predictive validity	Correctly predicts a specified outcome

2.2 The Information Pyramid

Indicators, and the data underlying them, can be considered in a hierarchical framework. Raw or primary data are indicators, although they may not convey much information on their own. As an example, vehicle fleet size does not necessarily present meaningful information unless combined with, for example, population data to produce a "motorization rate" often stated in vehicles per 1000 people. The "information pyramid" (see Figure 6-2) offers a simple way to understand the relationship between data, indicators and, ultimately, indices. Raw data form the foundation of the pyramid; this data can then be combined to construct different levels of indicators. Generally, movements up the pyramid are accompanied by a necessary simplification of the information and a loss of resolution. At the same time, however, if we consider the top of the pyramid to reflect the broader objective of interest (e.g. sustainable development), these movements up the pyramid can reflect a more easily understandable "indication" of achievement. Movement up the pyramid reduces the complexity of presenting a large range of data or indicators and can help screen out information. However, the level on the pyramid on which one focuses depends on the purpose of the indicators and data availability. Ultimately, one will face some degree of trade-off, such as between ease of comprehension and need for detailed resolution.

At the top of the pyramid is thc index, generally constructed mathematically from a number of contributing indicators. A well-known index is the *Human Development Index* (HDI), which the United Nations Development Program (UNDP) calculates as an attempted summary composite measure of human development for individual countries. The HDI aims to measure a country's achievements in three basic dimensions of human development (longevity, knowledge, living standards) using four indicators: life expectancy at birth, adult literacy rate, gross school enrollment ratio and GDP per capita (measured at purchasing power parity)

(UNDP 2001). The UNDP also calculates other national indices that represent for example, nations' gender equality, poverty levels and technological achievement (see UNDP 2001).

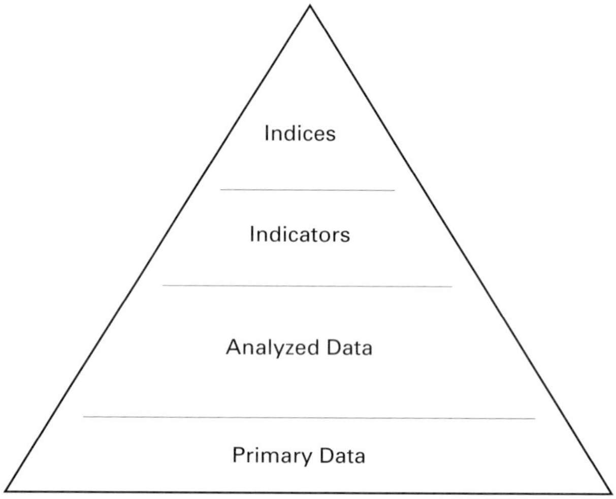

Figure 6-2. Data, indicators, indices: The information pyramid (Source: Spreng and Wils 1996)

While indices can prove useful in streamlining multifaceted and complex indicators into a single measurement, they also pose several challenges. First, an index itself does not explicitly show how its underlying components might be contributing to changes in the index value: a composite air quality index, for example, does not show the relative contribution of particulate matter versus ozone. Furthermore, the calculation of indexes ultimately requires a degree of subjectivity in weighting the contributing indicators. Transparency in the methodology used to calculate indices is, thus, critical to their effective use.

3. INDICATORS AND SUSTAINABILITY

Considerable work has been undertaken over the past 15 years to devise measures or indicators of the concept of sustainability. Nonetheless, an agreed upon standard set of "sustainability indicators" does not exist, largely because an operational definition of the sustainability concept remains elusive. Furthermore, the type of indicators to use depends on the purpose and the scale of analysis (e.g. national, regional, ecosystemic, etc.). Finally, there is the danger, mentioned above, that efforts in measurement, by nature, often lead us to measure what is measurable and ignore, by necessity, that which cannot be measured. In the case of sustainability or sustainable development, there is the risk that some of the most important aspects (e.g. intergenerational equity) are those for which little or no data exist. In this case, we may depend on the data least capable of actually evaluating the concept. This reality reinforces the need to adequately define the concept and, if no direct data exist, find valid and reliable proxies.

Table 6-2. Mapping the MAIN^{tetra} of Sustainability to the Four Types of Capital (Source: derived from Kain 2003)

MAIN^{tetra} dimension	Capital "component"
Mind	Human capital (e.g. ethics, worldviews, knowledge, skills and other human attributes)
Artifact	Human-made capital (e.g. works of art, instruments, machines, buildings and physical networks)
Institutions	Social capital (e.g. formal and informal networks, institutions, obligations (e.g., contracts), information systems, etc.)
Nature	Natural capital (e.g. air, water, soil, forests, mineral resources, "natural forces")

The economists' perspective offers one tractable approach to arrive at an operational definition of the sustainability concept. If we simply define sustainability as the capability to "maintain the capacity to provide non-declining well-being over time" (Neumayer 2003), then we can utilize the economist's perspective of maintaining the value of total capital, including human, natural, and manufactured capital (Neumayer 2003). (Note that these three "capital" components parallel the social, environmental and economic dimensions of sustainability.) Chapter 5, Section 5.2 introduces the MAIN^{tetra} (see Figure 5-5), which Kain (2003) explicitly links to the capital concept, more specifically to the World Bank's "capital stock" approach to sustainability. The MAIN^{tetra} is comprised of Man, Artifacts, Information, and Nature, which map to four types of capital: (1) human, (2) human-made, (3) social and (4) natural (Table 6-2). Kain (2003) suggests that the MAIN^{tetra} provides an analytical structure that could facilitate, among other things, the development and analysis of indicators. It should be noted, however, that the Tetra and its underlying capital forms do not provide an operational definition of sustainability on their own.

3.1 Sustainability Indices

If we start at the top of the information pyramid and return to the economist-oriented operational definition of sustainability from Neumayer (see above), then we arrive at sustainability indicators derived from traditional economic concepts, basically well-being or welfare and typical relevant and available indicators (or indices), such as GNP and GDP. Neumayer (2003) briefly traces the lineage of efforts to incorporate sustainability aspects into traditional welfare measures, starting with the *Measure of Economic Welfare* (MEW) in the early 1970s followed by the *Economic Aspects of Welfare* (EAW) in the 1980s and the *Total Incomes System of Accounts* (TISA) in the 1990s. Both the MEW and the EAW use basic techniques to try to account for the environment, in the former case, by adjusting for effects such as pollution and congestion and, in the latter case, reducing welfare by damage plus control costs (Neumayer 2003).

At the end of the 1980s, Daly and Cobb (1989) introduced the *Index of Sustainable Economic Welfare* (ISEW), also called the *Genuine Progress Indicator* (GPI). The ISEW basically attempts to adjust personal consumption (final demand or GDP) to account for income inequalities, social costs (e.g. crime, traffic accidents, commuting), environmental costs (e.g. air pollution, loss of environmental resources) and the value of welfare-increasing activities (e.g. housework, public infrastructure, net capital investment; Daly and Cobb 1989, Castañeda 1999, Neumayer 2000). ISEWs have been developed for several countries in Europe (including Germany, Italy, the Netherlands and the UK) as well as for Chile, with slight variations developed for Australia and the U.S. These various analyses come to a similar conclusion: At a certain point, GDP and ISEW diverge with GDP continuing to rise while ISEW stagnates or even falls. This is a phenomenon that Max-Neef (1995) points to as confirmation of his "threshold hypothesis," or

> *"For every society there seems to be a period in which economic growth (as conventionally measured) brings about an improvement in the quality of life, but only up to a point – the threshold point – beyond which, if there is more economic growth, quality of life may begin to deteriorate."*

Neumayer (2000) details some of the weaknesses in the assumptions underlying the ISEW calculations (e.g. the treatment of non-renewable resource depletion and the valuation of long-term environmental damage), which considerably weaken, though do not necessarily disprove, the threshold hypothesis. More recently, Neumayer (2003) focuses on a more fundamental critique: the difficulty in trying to combine the concept of well-being, which derives from the use of the current capital stock, together with the concept of sustainability, which relates to the value of the future capital stock into a single measure. To address this challenge, he proposes (Neumayer 2001) an indicator that combines the HDI with the measure of "genuine savings" (e.g. Hamilton and Clemens 1999), which have been estimated and published as "adjusted savings" by the World Bank since 1999 (see World Bank 2003). Basically, the genuine savings rate attempts to measure capital in terms of investment in and depletion of capital using, for example, education expenditures as a proxy for human capital investments and accounting for natural resource depletion. Currently, World Bank genuine savings rates calculated at the national level utilize very limited measures of environmental effects (e.g. covering a limited number of natural resource stocks and only including carbon dioxide as a pollution measure). The overall idea behind using the HDI together with genuine savings is to show the degree to which a society achieves current well-being (measured via the HDI) at the expense of total capital stock (measured by genuine savings).

Another interesting index, which approaches sustainability from a more ecological perspective, is the ecological footprint (Wackernagel and Rees 1996; see Chapter 4, Section 8.1 and Chapter 5, Section 2). While not typically characterized as an index per se, the ecological footprint essentially functions as an index by attempting to measure consumption and waste production and converting the resulting resource flows into an estimate of the biologically productive area needed to provide these functions (Wackernagel and Rees 1996). The overall goal is to present a measure of ecological sustainability in an understandable and comparable way: in terms of the per capita land area required. In other words, the ecological footprint can be calculated in an attempt to show how much land would be required to support the consumption of the average inhabitant of a given country. Ecological footprints have been calculated for many countries (e.g. Mexico, Switzerland, Canada, India, Chile) and cities. Levett (1998) offers some insightful observations and advice on the use of the ecological footprint.

Indicators such as those proposed by Neumayer operate on a useful scale for international comparisons. However, they naturally suffer from lack of data (e.g. genuine savings rates are not available for all countries) and incomplete measurement (such as limited representation of resource stocks and pollutants). Furthermore, questions remain regarding their practical application, particularly in the urban development sector: can they effectively guide policies or aid in management? How well can they serve to measure progress and guide changes?

3.2 Sustainability Indicators

Movement down the information pyramid (Figure 6-2) leads to a natural increase in the number of indicators that can be used to measure the sustainability concept. For example, as part of on ongoing program to develop national-level indicators, the United Nations Division of Sustainable

Development, through an iterative process of testing and improvement, currently proposes an indicator framework including 58 different indicators, representing various themes and sub-themes in four different dimensions of sustainable development (see Table 6-3).

Table 6-3. UN Framework for Sustainability Indicators (Source: UN DSD 2004)

Dimension	Theme (sub themes)	Indicators
Social	Equity (poverty, gender equality)	4
	Health (nutritional status, mortality, sanitation, drinking water, healthcare delivery)	8
	Education (educational level, literacy)	3
	Housing (living conditions)	1
	Security (crime)	1
	Population (population change)	2
Environmental	Atmosphere (climate change, ozone depletion, air quality)	3
	Land (agriculture, forests, desertification, urbanization)	7
	Oceans, seas and coasts (coastal zone, fisheries)	3
	Fresh water (water quantity, water quality)	3
	Biodiversity (ecosystem, species)	3
Economic	Economic structure (economic performance, trade, financial status)	5
	Consumption and production patterns (material consumption, energy use, waste generation and management, transportation)	9
Institutional	Institutional framework (strategic implementation of SD, international cooperation)	2
	Institutional capacity (information access, communication infrastructure, science and technology, disaster preparedness and response)	4

Note that these four dimensions map exactly to the MAIN^{tetra} introduced in Chapter 5, Section 5.2.

While sustainability indicators often tend to be presented in the three (or four) dimensions of sustainable development (see Table 6-3), examples of other indicator categorizations exist. For example, Portney (2003) reports on Colson's five categories of indicators: (1) resources (ecological, environmental, natural); (2) economic performance and equity; (3) ethical considerations; (4) sociocultural issues and (5) political governmental functions.

Returning to the issue of system complexity and feedback introduces the practical question of whether or not we can come up with indicators that not only track changes in the whole system and its subsystems (e.g. Hodge et al 1999) but can also account for the interactions between the subsystems themselves. Gustavson et al (1999) echo this concern, noting that indicators dealing with the three sustainability dimensions are "usually addressed in isolation of each other." On the other hand, indicators may be overlapping, e.g. "air quality" may refer to the environmental as well as to the social dimension (health) of sustainability. This raises the need to further differentiate between indicator types. The *Pressure-State-Response* (PSR) framework (see e.g. OECD 1997) offers one means of differentiating between various indicators: (1) pressures or stresses, which represent impacts on systems (e.g. congestion, pollutant emissions); (2) states, which represent conditions over time (although these should be distinguished from pressure indicators often used as proxies for state indicators, see OECD 1997); and (3) responses, which generally aim to reflect efforts to respond to pressures. The European Environment Agency has extended the PSR framework to include impacts (which precede responses) and driving forces (which precede pressures and are also affected by responses). The DPSIR (driving forces-pressures-states-impacts-responses) thus forms a closed loop analytical structure, with indicators possible within each area (see e.g. EEA 2002 and Figure 6-3).

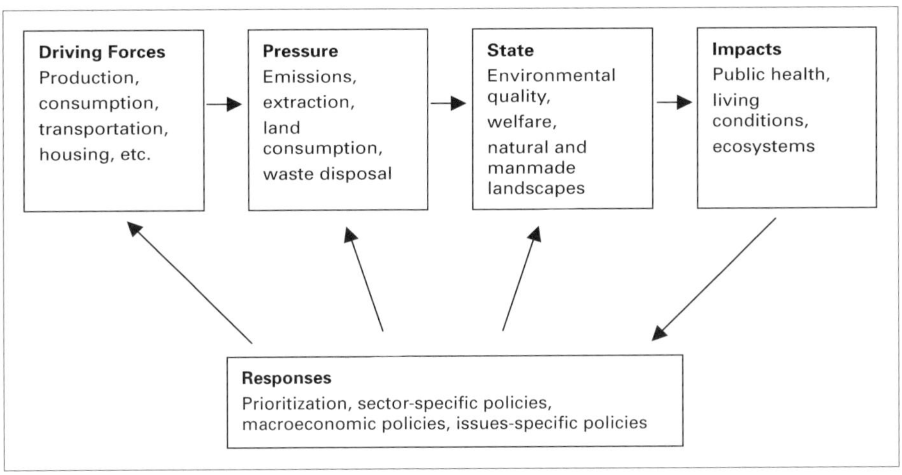

Figure 6-3. The DPSIR Schematic (Source: EEA 2002)

Beyond the criticism of indicators being addressed in isolation, Gustavson et al. (1999) raise two additional criticisms: the fact that indicators are often developed along political boundary lines, while ecosystems rarely, if ever, match such boundaries (see discussion in Section 1.1), and that work on sustainability indicators often lacks a "projective perspective" (i.e. with aims to predict future conditions). Finally, it is important to reiterate that we cannot ignore important issues simply because they are not measurable by a valid and reliable indicator and that measurement has an important role in helping to understand both quantitative and qualitative aspects (Hodge et al. 1999).

3.3 Urban Indicators and Indicators of Sustainable Urban Development

Of course, the use of indicators in urban planning is not new. For example in 1945, the American Public Health Association (APHA) released a publication aimed at improving and making uniform the methods used to measure housing quality (e.g. Hodge 1963). In 1978, the OECD published a report identifying a core set of housing indicators: indoor space, outdoor space, amenities and sanitation, security of tenure, housing cost and access, and services and employment (OECD 1978). Since the late 1980s, performance measures have also formed an integral part of housing management in the United Kingdom (Smith and Walker 1994).

The OECD indicators from 1978 can be seen as something of a direct predecessor to urban sustainability indicators. Building on this precedent, the European Union further developed a set of indicators for the urban environment through its Environment Agency (EEA), initially identifying 55 indicators classified under 16 different attributes in 3 general areas (see Table 6-4). While intended to enable a comparison of conditions among 72 European cities, the lack of comparable data meant that only 20 indicators could be assessed systematically across just 51 of the cities (OECD 1997). The UN HABITAT (2004) encourages the development of quantitative and qualitative indicators to help "monitor global progress in implementing the Habitat Agenda and to monitor and evaluate global urban conditions and trends." In this regard, HABITAT has developed 23 key indicators, together with areas for qualitative information, intended to help measure city performance in implementing the Habitat Agenda. HABITAT has assembled a larger database of indicators (90 quantitative indicators and 35 qualitative indicators) in an attempt to measure these for 232 cities around the world (UN HABITAT 1998). If the EEA indicators project was unable to find comparable data for just 72 European cities, one can only imagine the difficulty of compiling data for 232 cities around the world. Overall, an average of just half of the 90 quantitative indicators are available for the cities in the HABITAT database.

Table 6-4. EEA Selected Urban Environment Indicator Areas and Attributes (Source: OECD 1997)

Area	Attributes	Sub-attributes
Indicators of urban patterns	Urban population	Population, population density
	Urban land cover	Total area, total built-up area, open area, transportation network
	Derelict Areas	Total area
	Urban renewal areas	Total area
	Urban mobility	Mode split, commuting patterns, traffic volumes
Indicators of urban flows	Water	Water consumption, wastewater
	Energy	Energy consumption, energy production plants
	Materials and products	Goods transport
Indicators of urban environmental quality	Waste	Waste production, recycling, waste treatment
	Water quality	Drinking water, surface water
	Air quality	Long term and concentrations
	Acoustic quality	Noise exposure
	Traffic safety	Fatalities and casualties
	Housing quality	Average floor area per person
	Greenspace accessibility	Proximity to urban green areas
	Quality of urban wildlife	Number of bird species

The European Commission also initiated the European Sustainability Index Project in 1993, which identified ten issues that should constitute a Sustainability Index for cities: healthy environment, green space, efficient use of resources, quality of built environment, accessibility, green economy, vitality, community involvement, social justice, and well-being (OECD 1997). As mentioned in the previous section, the ecological footprint can be considered as a form of sustainability index, at least in the environmental or ecological sense. Ecological footprints have already been calculated for many urban areas, including Santiago de Chile (Wackernagel 1998; see Section 3.9 of this chapter).

In the United States, Portney (2003) notes that many local governments, NGOs, and private companies have worked to develop sustainability indicators, generally building upon existing data and covering a range of relevant issues. Among the most often cited cases (e.g. Portney 2003, Newman and Kenworthy 1999, OECD 1997) is the Sustainable Seattle initiative, which includes some 40 indicators categorized by environment, population and resources, economy and culture and society (Sustainable Seattle 1998). According to Portney (2003), the multidimensional approach to sustainability has taken hold in local initiatives to develop indicators in the U.S. However, he feels that there has been a lack of discussion of why these indicators are being chosen. Portney ultimately questions whether this "new paradigm" of indicators has actually translated into indicator practice, which he maintains remains wed to traditional city functions and areas of activity (both due to data availability realities and the need for measures of performance for government functions). Over the last few years, the U.S. Environmental Protection Agency (EPA) has contracted the development of its *Smart Growth Index* (SGI), which is a GIS-based software that uses 56 indicators as benchmarks to help communities evaluate impacts of alternative land use plans (US EPA 2002). While not explicitly aimed at measuring sustainability, the eight categories of indicators – land use, housing, employment, environment, parks, infrastructure, transportation, air quality and climate change – do map relatively closely to sustainability "dimensions."

In the developing world, relevant recent sustainable indicator work includes an analysis of the Guongzhou urban area, which included the development of a sustainability index and an indicator of coordination among the three sustainability dimensions (Weng and Yang 2003). At a regional level, Herrera-Ulloa et al. (2003) use principal component analysis to derive a sustainability index from 27 indicators in the four sustainability dimensions (including institutions) for the Baja California region of Mexico.

3.4 The Sustainable Indicator Prism

The broad range of characteristics to be measured for a comprehensive view of SUD can easily lead one, whether policymaker, citizen, or business owner, to become quickly mired in "information overload." Using a large battery of indicators to measure a wide variety of phenomenon can potentially make us lose sight of the final goal, drowning our analysis in a vague representation of extensive information with insufficient focus. Indicators, in this case, become an unmanageable list of tasks. Structuring sustainability indicators into the information pyramid (Figure 6-2) offers one way to help overcome this problem. Building on the information pyramid and the MAIN^{tetra} (Figure 5-5), we propose the sustainable indicator prism as a way of conceptually structuring the hierarchy of information implicit in sustainability indicators (see Figure 6-4). Each side of the prism represents one of the sustainability dimensions: data forms the base or foundation of the prism, and movement up the sides of the prism leads to a consolidation of information into different tiers of indicators and, ultimately, a measure of broader achievement. Figure 6-5 presents a detailed illustration of the envisioned process.

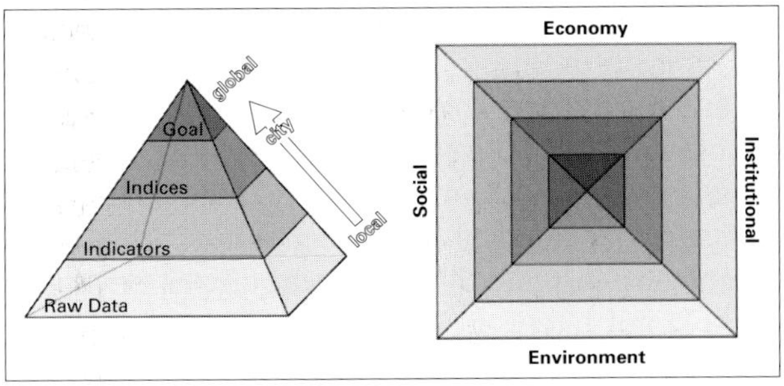

Figure 6-4. The sustainable indicator prism

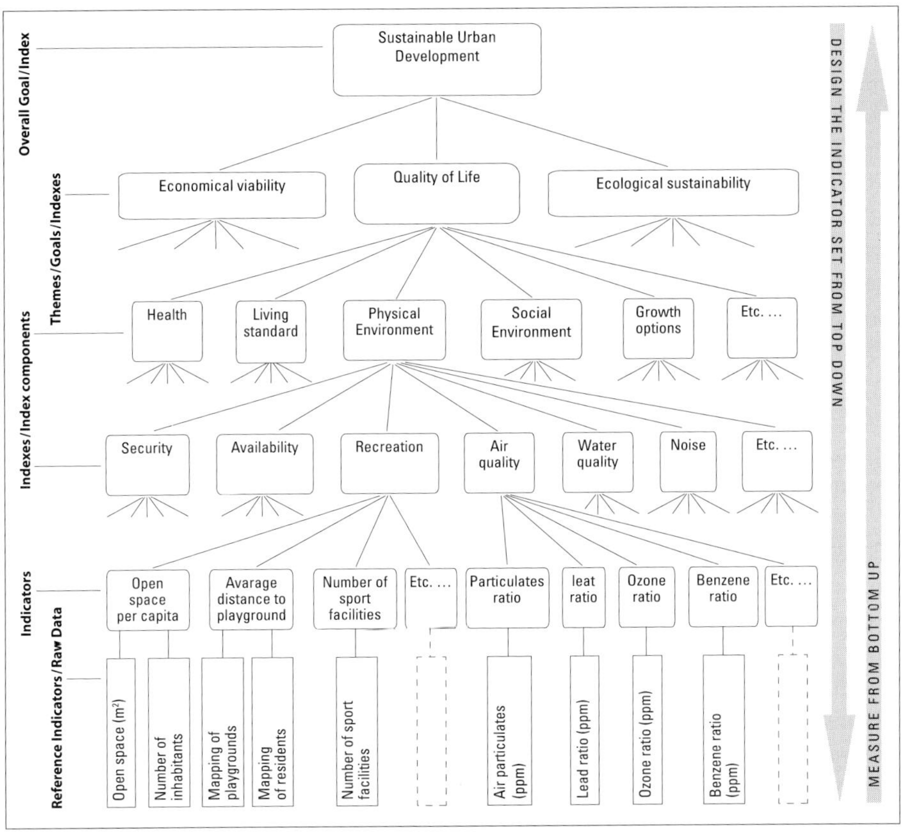

Figure 6-5. Sustainable indicator hierarchy: From data to indicators to indices (Source: inspired by Mitchell 2000)

3.5 Indicators of SUD and the Scale and Context of Assessment

Figures 6-4 and 6-5 convey a sense of "informational scale." In the urban setting, the geographic or physical scale is also relevant to indicator development and use. For example at the metropolitan level, the UN HABITAT indicators, discussed above, monitor urban conditions at an international comparative scale. For the local citizen or municipal manager, such indicators will not likely provide the resolution necessary to assess the condition of interest. While some concepts, such as poverty, can be measured from the family all the way to the global level, other concepts are unique to a location or have different relevancies depending on the setting. For example, Bossel (1999) lists crime rate and the number of children in poverty as useful indicators for the country of New Zealand. Such indicators have the advantage of being easily comparable to other cities and are generally appropriate measures for any place. On the other hand, the relevant indicators necessarily have to vary at the neighborhood level. Students from the Santiago Urban Planning Studio, for example, identified the percentage of the population that felt comfortable walking home at night in a certain area as an appropriate indicator to gauge the effects of their proposed streetscape improvement ideas (MIT-CDD 2003). While it is feasible at the neighborhood level to interview people on their perceptions, this level of detail is not always possible as scale and subject population increase. Hemphill et al. (2004b) tackled this problematic by developing 52 indicators within five different groups (economy and work, resource use, buildings and land use, transport and mobility and community benefits) to measure the relative sustainability of the different developments of various site-specific urban regeneration schemes in three European cities.

While Hemphill et al. (2004a,b) imply some degree of uniformity in assessing sustainability at the site-specific scale, it may still be necessary to use measures that reflect very specific local qualities of importance. For example, Bossel (1999) includes a count of wild salmon as an indicator in his set for the city of Seattle. These types of indicators add a level of uniqueness and richness that describe a specific place. Special concerns and issues can also be addressed through indicators that are place-oriented. The ethnic mix of a neighborhood and relative wealth of each group are particularly appropriate indicators for post-apartheid Johannesburg, while these same indicators do not have the same meaning or importance in Santiago. This relates again to the ultimate purpose of the indicators, whether they are to be used for local performance measurement or intermetropolitan comparisons. Again, drawing from the example of the Santiago Planning Studio, students identified one potential indicator as the amount of new investment within a 500 meter radius of a transit-oriented development project. While this indicator may be useful in evaluating the effects of this specific project, it may not serve to compare this project to another transit-oriented development project in, for example, Gaborone. Each project is unique and may well have a different objective, extent of investment and urban context.

3.6 Producers and Users of SUD Indicators

At the risk of further complication, we need to point out another important dimension related to indicators of sustainable urban development: stakeholders, or more simply, the people who ultimately produce and use indicators. As referred to above, some international organizations (e.g. UN-HABITAT) are interested in indicators to assess the "state" of the world's cities; we might call this a measure of the indicators' "comparative utility." At the local level, however, interest in indicators may be more mundane, yet ultimately more germane. Local residents, for example, may naturally be most interested in indicators that measure the internal health of their neighborhood and may most care about methods for measuring the effect of specific projects in their neighborhoods. We can call this a measure of indicators' "internal utility." Of course, comparative indicators will also be useful to local residents to give them an idea of how their neighborhood compares to other communities.

We can conceptualize at least three general interest groups as users of indicators at the city level: local residents, the private sector and the municipal government. Each group might be interested in indicators of different scales or utilities, while some indicators may be universally important. This reflects the inherent political context of indicators. The use of performance measurement indicators provides citizens a way to monitor the effectiveness of government policies and actions (Innes and Booher 1999), which can keep the government accountable for its decisions. Ideally, the government is interested in satisfying the demands of local residents while also presenting information about the community that appeals to outside investors. Therefore, the government's set of indicators would tend to encompass the residents' sets while also incorporating new indicators that allow for further comparison with other communities.

The private sector, especially those looking to invest in a community, are most interested in the unique qualities and internal health of a neighborhood as well as its competitiveness with other areas. For example, a real estate developer would look at what land uses a community already supports and where potential new markets may exist. However, comparability becomes important when many opportunities for investment are available. In some cases, the interests of the private sector, the local community and the government may converge; in others, however, they may not.

In recognizing the relationship between indicator and "user," Innes and Booher (1999) highlight the need for three types of indicators:

- System indicators: a few key measures that reflect central values of concern (these might be considered indices)
- Performance measures: specific outcomes of various system aspects, such as the street system, or social services provision, which can be directed towards policymakers and planners
- Rapid feedback indicators: to help individuals, decision makers, businesses make daily decisions in "real time"

This last indicator type, the "rapid feedback" indicator, reminds us again of the dynamism underlying the sustainability concept: we are agents in the system, our actions affect the system and the information we discern from the system (via indicators) influences our actions within it.

3.7 Participation and Politics: Stakeholders and SUD Indicators

Innes and Booher (1999) believe that engaging the stakeholders is more important than "getting the indicators exactly right," and indicators should be used as a tool for discussion among all interest groups in the planning process. In a study of indicator sets in four western European case cities, the Pastille Project Consortium concluded that indicators

> *"[...] can help organisations assimilate and better understand stakeholder views regarding sustainable development."* (PPC 2002)

The role of participation in indicator development, is not, however, entirely clear-cut. Brugmann (1997a,b), for example, considers that indicators at the local level are suboptimal tools for public education, due to the technical complexity of the systems of interest (see e.g. Section 1.1 of this Chapter). From this perspective, an underlying tension exists between the need for "scientific rigour" and the need to account for "public values and perceptions" (Brugmann 1997a). In assessing the Sustainable Seattle project, Brugmann concludes that the initiative was a valuable "public education and advocacy exercise" but falls short as a serious assessment of sustainability, which would require a more "extensive and technically informed assessment effort" (1997a). The indicator work of Hemphill et al. (2004a) offers an example of this perspective, as they utilized a Delphi method and multicriteria analysis to develop weighted indicator groups and relevant indicators based on iterative consultations with *expert* panels of academics and professionals. Weng and Yang (2003) offer another example of the "expert/scientific approach" to indicator development in the urban setting.

On the other hand, however, many make a strong call for active participation in indicator development (e.g. Innes and Booher 1999, Pinfield 1997). Leitman (1999), though specifically focusing on "Quality of Life" (QOL) indicators as a tool for sustaining urban development, places heavy emphasis on the participation of relevant groups (experts, decision makers, and stakeholders). He notes that indicators should be developed with participation from groups representing the same level at which the indicators will ultimately be applied (Leitman 1999), which suggests that experts, *local* decision makers, *local* residents and other stakeholders should work together to develop a relevant set of indicators that suits the needs of the *local* community and its scale. Even with this perspective, however, public participation poses challenges. One concern is whether all stakeholders have been represented in the indicator development process. Poorer communities may be limited in their capacity to participate because of a lack of basic skills or time. In addition, experts may exert too much bias in the indicator development process, yielding a set that is not truly representative (Leitman 1999).

This latter point highlights the inherently political nature of sustainability. Crilly et al. (1999) go so far as to explicitly call sustainability a "political" and not "technical" issue, considering the political process as the only way to incorporate community values and goals into an indicator set that will ultimately represent the concerns of all stakeholders. In this regard, Crilly et al. (1999) make special mention of the potential value of nongovernmental organizations as a means of providing information to disenfranchised groups and serving as a communication line between the general public and the government.

Where, however, do these somewhat diverging views leave us in terms of the role of participation in indicator development? The use of the local *Integrated Development Plan* (IDP) in Johannesburg (see Chapter 5, Section 3.2) provides one good example. In this process, stakeholders are brought together in all eleven of Johannesburg's municipal regions. The IDP serves as a comprehensive plan for each municipality at the local level and is required to align with provincial and national plans. It is also designed to link the municipal government with the other two interest groups, local residents and businesses. Each local IDP must include a list of key performance indicators. The specialized knowledge of experts is used to develop indicators once the values and objectives of the stakeholders have been established. Finally, the local plans are combined to create an integrated development plan for the City of Johannesburg as a whole.

3.8 Data: Producers and Users

In the end, of course, the process of SUD indicator development cannot occur outside the political realm – an obvious point, perhaps, but one worth bearing in mind. Even the data underlying indicators arise from political pragmatism relating to institutional responsibilities and agendas. Issues or concerns with no strong historical political constituencies may well be immeasurable, if not due to their inherently complicated nature (as discussed earlier) then simply because the relevant data does not exist. Industrialized world studies highlight the problem of data availability (e.g. Gustavson et al 1999, OECD 1997); in the developing world, the challenges are generally more severe.

In essence, data is "produced" by all of the various groups of actors in the urban area: the public, the government, and the private sector. After all, our combined actions produce the housing prices, traffic flows, pollutant emissions, etc. underlying all data. The challenge, however, comes from the processes of systematizing the collection, storage, and ultimate access to and use of the data for the purposes at hand. The government is often the responsible party for such activities and yet governments are often time and resource-constrained, making them unwilling to make the relatively modest investment necessary for the generation of adequate data. Even when the government engages in consistent data collection, such activities often occur in a compartmentalized fashion, with data collection logically matching institutional responsibilities and often differing in terms of time periods of collection and degree of spatial disaggregation.

For example, in the Santiago Metropolitan Area, at least 11 different public agencies can be identified as collecting data relevant to the construction of sustainability indicators (see Table 6-5). The Table distinguishes between how the data is typically presented (i.e., spatial scale of aggregation) and how it is generally made accessible for use by others. As can be seen, data is not always collected at the same spatial scale and/or not always geo-coded, which may not be an issue in the creation of metropolitan-wide indicators, but certainly presents a challenge to the development of scalable and scale-consistent indicators. For Greater Santiago, for example, a household travel (origin-destination) survey conducted in 1977, 1991 and 2001, offers a rich degree of information not only on travel characteristics but also other socio-economic and demographic information, such as household size, income, job and education status, etc. At the raw level, the travel survey data is now geo-coded at the level of specific addresses; however, the information is often made available at more aggregate levels (such as traffic analysis zones for use in travel forecasting models), which are then not coherent with the units of spatial analysis defined, for example, in the Population Census. This makes it difficult to bring these two potent data sources to bear in indicator development.

Table 6-5. Various potential data sources for sustainability indicators in Santiago de Chile

Source	Types of data	Geographic level	Access
INE	(1) Population (census) (2) Building construction (permits) (3) Transportation (i.e., vehicle registrations) (4) Socioeconomic conditions	(1) National aggregated data (regional, municipal, census tract, block, parcel) (2) Municipal level (3) Regional level (4) Municipal level	(1) Detailed at household level (2) Project information available at municipal government (3) Only aggregated data (4) Municipal level
MIDEPLAN	Socioeconomic survey (semi-annual): household spending, income, etc.	Municipal	Municipal level
SII	Real estate activities (residential, commercial, industrial, etc.)	Parcel	Limited access due to confidentiality
SECTRA	(1) Origin-destination surveys, (2) Traffic counts and intercept (cordon) surveys (3) Vehicle emissions factors	Metropolitan and municipal	(1) Municipal, TAZs, blocks (2) local counts (3) general data
CONASET	Traffic accidents, incl. mode, location, type, cause, etc.	National and local	In Santiago, geo-referenced
CONAMA	General data on environmental conditions	National and regional	General data
SESMA	Air pollution, noise pollution data	Metropolitan region, monitoring sites	Hourly/daily pollution readings for air pollution
MINVU	Constructed/approved public housing	Lot, census tract, municipal	Available at Municipalities
MOPTT	Infrastructure investments, bus concessions, operations	Regional	Reports, internal documents
SAG	Agricultural land use, changes	Unclear	Reports
Metro	Ridership, investment, rolling stock, operating costs (including energy use, depreciation)	Stations, lines	Annual reports
Private companies	Water consumption, waste disposal, electricity	Consumer, service area	General data

Notes: INE: National Institute of Statistics; MIDEPLAN: Ministry of Planning; SII: Internal Tax Service; SECTRA: Inter-ministerial Secretary for Transportation Infrastructure Planning; CONASET: National Commission for Traffic Safety; CONAMA: National Environment Commission; SESMA: Metropolitan Health and Environment Service; MINVU: Ministry of Housing and Urban Development; MOPTT: Ministry of Public Works, Transportation and Telecoms; SAG: Agricultural and Cattle Service; Metro: subway operating company; TAZ: Traffic Analysis Zones

Theoretically, the process of moving from raw data (information production) at the base of the Indicator Prism (Figure 6-4) towards indicators can result in a potential loss of important information. In practice, however, due to the process of data collection and storage, data tends to become less reliable as the unit of spatial analysis shrinks (see e.g. Gustavson et al. 1999), reflecting the fact that data is generally made available in a relatively aggregate form, from which disaggregation must then take place. Further challenges arise from the fact that many government agencies keep data internal ("information is power"), thereby further inhibiting indicator construction. Sometimes, specific statutes govern this data availability. Given relative data scarcity, even government agencies often consider their commercial value so that the private sector or even other government agencies must pay for access. When the private sector specifically collects data, the reasons for limiting access by other parties are obvious: even when others are willing to pay, recognizing data as a source of competitive advantage may preclude the sale of relevant information. Advances in information and communications technologies hold some promise for data development and management. Cellular phones with global positioning systems, for example, can be used to provide real time information on traffic systems; computing advances allow for storage, management and manipulation of massive amounts of data, which can also be represented easily in Geographic Information Systems (GIS). Other sources show promise in revealing information that could otherwise take years of painstaking fieldwork to collect. For example, Frazier and Kockelman (2004) demonstrate initial steps in utilizing "the skies and their satellites" to interpret the texture and pattern of Austin, Texas (USA). Combining land use cover data from satellite images and cartographic maps, they compute a variety of spatial attributes, including land use mix, with good results.

3.9 Existing Indicator Sets for Johannesburg, Gaborone and Santiago

The indicator sets available for each of the three cities under study vary widely. Either in their systematic approach, their purpose or in their source, the cities have adjusted the groups of indicators to their specific needs and have identified individual problems and intervention objectives.

In the case of Johannesburg, there are families of indicators associated with diagnoses and plans that cover a wide range of goals. Johannesburg's "balanced city scorecard approach," with its mandatory requirement to include key performance indicators for each local *Integrated Development Plan* (IDP), illustrates a comprehensive indicator development process. An attempt is made to approach the planning process from a grassroots, bottom-up perspective where public participation plays a strong role. Indicator development begins at the local scale (with the IDP) but is incorporated into the citywide plan. However, the current bias in Johannesburg's indicator set is towards the equity and economic components of sustainability, with environmental concerns appearing low on the priority list.

Botswana has also been able to propose extensive sets of indicators on the national level, although their application is less explicit. Three subsets of indicators are currently being used. The first is based on the United Nations *Agenda 21* for Botswana. The second, entitled *Vision 2016,* is a strategic framework supported by the United Nations Development Programme, which includes an online monitoring and evaluation database with over 100 indicators. Finally, on the urban level, a third set developed by the United Nations Center for Human Settlements provides 23 quantitative indicators designed to allow Gaborone to compare its progress towards sustainability with Francistown, the second largest city in Botswana. Despite these indicator sets, the City of Gaborone still lacks a single comprehensive one. Bührer (2002), building on the existing local and national indicator sets, proposes 30 indicators that he judges to be effective in describing Gaborone's progress towards sustainability. Based on his judgment, the proposed indicators differ relative to developed world indicator sets in terms of their relative treatment of education, health, informal employment and infrastructure provision. Bührer (2002) discusses in detail the possibilities for (and difficulties with) actually developing the proposed indicators based on available data in Gaborone.

In Chile, no official sustainability indicators of currently exist for the city. In 1997, the National Environment Commission initiated a project to develop indicators of sustainable development for each of Chile's 13 regions, including the Metropolitan Region of Santiago or RMS (Quiroga 2001). The *National System for Environmental Information* (SiNIA) currently administers the sustainable indicators at the national and regional level. For the RMS, there are 14 indicators all with a heavy environmental focus (SiNIA 2004). There has also been an attempt to systematize territorial data during the last years through the creation of the *National System of Territorial Information* (SNIT) in 2001, under the auspices of the interministerial Commission on the City and Territory and managed by the Ministry of Housing and Urban Development (MINVU). The initiative aims to generate an inventory of existing territorial information, create norms and standards for information collection and elaborate a national plan for data collection and standardization (SNIT 2003). While not specifically geared towards sustainability indicators, the initiative could, in theory, lead to information more conducive to indicator development. More recently,

MINVU has also launched an Urban Observatory, with the goal of developing indicators of urban performance across the country.

More generally, it is quite typical for different government agencies to utilize traditional performance measures in Chile. The Ministry of Housing and Urban Development (MINVU), for example, measures progress, in part, by the number of affordable housing units built annually. The Corporation for the Development of Santiago (a quasi-governmental urban renewal agency) measures progress, in part, based on the number of urban renewal subsidies issued per year. One example of environmental indicator use is the declaration of "environmental emergency days," based on air pollution monitoring measurements as well as predictive pollutant dispersion models. On such days, certain activities are curtailed, depending on the degree of the problem, including restrictions on vehicle travel (*La Restricción*). Municipal governments use data available through national sources (e.g. Population Census and semiannual socioeconomic surveys) in order to assign and prioritize social investments, and the Central Government uses similar data sources to justify the redistribution of resources to Municipalities via relevant transfer mechanisms.

Outside official government efforts, Wackernagel (1998) applied the ecological footprint methodology (as mentioned in Section 3.3) to the Santiago metropolitan area, aiming to provide a measure of the ecological capacity used by the city to sustain its functioning (i.e., its dependence on natural capital). Based on an admittedly rough and preliminary analysis, Wackernagel (1998) estimates that the average Santiago resident requires 2.6 hectares per capita to sustain his/her consumption.

3.10 Developing Comprehensive Sustainability Indicator Sets

Ultimately, the determining of a group of indicators is intimately associated with each city's particular sustainability goals and challenges. Some indicators serve a comparative utility by allowing each city to measure its progress against others. These types of indicators will be part of every city's comprehensive set, including economic indicators such as employment levels, social indicators dealing with poverty, crime, and mortality levels and environmental indicators such as level of urban expansion and pollution. However, many indicators aim towards specific goals or allow for performance measurement and the internal evaluation of projects and programs.

The first step in the development of a comprehensive indicator set is to clearly recognize the citizen's values and establish goals for sustainability. At this stage, perhaps, the clearest role for explicit public participation may exist. The actual selection of indicators should be a joint process undertaken by the public (laypersons), the private sector, policymakers and experts. While the beginning of the indicator development process can take a bottom-up approach with the involvement of laypersons, scientific and expert opinion is necessary at the top. The key lies in effectively linking the two approaches, a highly problematic task, but one that is critical, not only to indicator development but to the implementation of sustainability strategies (Selman and Parker 1997). Perhaps the most important strategic detail in the process is the development of an operational definition of sustainability at the relevant scale of interest.

In their effort to develop indicators of and model sustainability in the broad Fraser River Basin (a large ecosystem area in northwest Canada, including the Vancouver metropolitan area), Gustavson et al. (1999) offer useful lessons and cautions to indicator use. Those authors suggest that focusing on a small number of indicators within selected classes (e.g. economic, social, environmental or human health; see also the sustainable indicator prism in Figure 6-4) provides a fruitful and productive approach, since the often close correlation among many indicators in relevant classes means that a larger number of indicators will not necessarily improve policy guidance. Interestingly, and with this policy focus in mind (linked, also, to the DPSIR framework in Figure 6-3), Gustavson et al also suggest that precision in indicator estimation is often overly emphasized, particularly since it is almost impossible to achieve for many relevant issues. This leads to another lesson of Gustavson et al. (1999), that the selection (and modeling) of sustainable development indicators is more a "judgmental art" than an "exact science" (echoing Meier and Brudney 2002, as discussed in Section 2.1).

In fine-tuning sustainable indicator development, one must keep in mind the ultimate relevance to project objectives (related to spatial and temporal scales as well), links to the underlying cause-effect processes and of course realistic expectations in terms of the availability and/or costs of data collection and systematization (Segnestam 1999). We have suggested in this Chapter that although a comprehensive urban indicator set might be useful for intercity comparisons ("comparative utility") and general accountability for citywide managers and politicians, we also expect much more variability in the actual types of indicators depending on their purpose, for example, to analyze a brownfield redevelopment or neighborhood revitalization ("internal utility"). Even at the citywide level, the relevant indicators may vary depending on the context and objectives, whether they are competitiveness (Johannesburg), control of urban expansion (Gaborone) or the mitigation of the environmental impacts of expansion (Santiago).

Ultimately, a comprehensive indicator set must have a real and concrete purpose.

In a strongly cautionary take on the "appraisal" of sustainability, Stirling (1999) offers some useful observations related to the use and development of indicators, including the need to recognize sustainability as a process with inherent subjectivities and uncertainties. From this assessment comes several suggestions relevant to indicator development: transparency, disaggregate treatment of the different sustainability dimensions and a focus on sensitive or crucial issues. With such cautionary words in mind, we again suggest that the sustainable indicator prism can offer a framework for constructing indicators and that these indicators should focus on the current estimated "state" of the condition (e.g. education level), its direction over time (e.g. remaining flat) and a desired directionality (e.g. increasing).

4. CONCLUSIONS

We live in an age of instant information where the public increasingly views access to information as a key to their ability to understand and make the right decision. But as has been pointed out in this chapter, there is no guarantee that the quality, accuracy or scale of the information is sufficient to develop common agreed upon Sustainable Urban Development indicators. There is also no guarantee that the data used and presented is relevant without establishing its appropriate context, scale and geo-reference.

Much has been written, discussed and debated about how urban development should be measured and what the correct indicators for assessing its level of sustainability could be. Whatever opinion one may have, there is still agreement that indicators, particularly those that attempt to measure Sustainable Urban Development, are still useful:

- In raising awareness of development issues among all sectors of society
- In facilitating the understanding and definition of urban sustainability
- In helping in assessing the state of our own cities (or neighborhoods) – what we have termed "internal utility"
- In providing a standard mechanism for comparison with other cities (or neighborhoods) – what we have termed "comparative utility"
- In pointing to the many challenges and limitations associated with data gathering and information sharing

Sustainability indicators do not just measure change. They also describe the direction a place or a community may be heading toward in a way that is easily understandable. By illuminating the spectrum of relevant issues, indicators link various elements that are often seen and discussed separately (housing, ecology, safety, etc.). In the right context and setting, indicators not only help the public evaluate whether actions are having the desired effect, but also facilitate in determining what conditions exist and whether the direction is consistent with stated goals.

By raising the awareness of the state of affairs and the condition of a place, indicators allow a community to hold itself, its public officials, and supporting institutions accountable, as well as help to build a consensus for future actions. In order to achieve such a consensus, it is necessary for partnerships to be developed between various sectors. With an increasing recognition that no single sector or institution is capable of making all the changes necessary or solving all the sustainability challenges confronting society, such partnerships must be made at all contextual levels to integrate decision-making among national, state, and local jurisdictions. Effort must also be made to overcome the barriers between competitors, regulators as well as community stakeholders, because these institutional barriers also form informational (data) barriers.

Beyond raising awareness and overcoming institutional barriers, the development of indicators is dependent on accurate and readily available data. In recent years, the volume and detail of spatial data has exploded and the GIS technologies for manipulating the data have become much more capable and standardized. However, many data integration issues remain unresolved as they involve difficult generalization and compilation problems that are hard to automate. The creation of a more coherent yet flexible spatial information infrastructure will give any entity dealing with place-based problems and urban development an improved mechanism for tapping into the types of information that have been difficult to share. Far more data must become usable and accessible via a broad range of user interfaces that would simplify the development of Sustainable Urban Development indicators across many domains.

Finally, and most importantly, valid and reliable sustainability indicators can be created only through some form of community involvement in the process. Without the engagement of a diverse and representative set of community stakeholders it would be difficult to reach a long-term vision for it's the community's future. Without such a vision, goals cannot be categorized and indicators related to these goals cannot be developed.

This process of developing and using indicators works best when communities are in command of the process, drawing upon professional expertise when appropriate. Experience has shown that when professionals try to dominate, selecting indicators often turns into an academic exercise devoid of reference to local experience and place. A balance between the two must be struck, one that distinguishes and codifies place-based urban assets and local wisdom while following appropriate professional standards and scientific rigor.

REFERENCES

Bossel, H. 1999. *Indicators for Sustainable Development: Theory, Method, Applications*. A Report to the Balaton Group. Winnipeg: International Institute for Sustainable Development.

Brugmann, J. 1997a. Is There a Method in Our Measurement? The Use of Indicators in Local Sustainable Development Planning. *Local Environment*, Vol. 2, No. 1: 59–72.

Brugmann, J. 1997b. Sustainability Indicators Revisited: Getting from Political Objectives to Performance Outcomes – a Response to Graham Pinfield. *Local Environment*, Vol. 2, No. 3: 299–302.

Bührer, C. 2002. *Indicators in Sustainable and Integrated Planning in Gaborone*. Diploma Thesis. Institute for Territorial Development and Landscape, Department of Civil, Environmental and Geomatics Engineering, Swiss Federal Institute of Technology, Zurich.

Castañeda, B. 1999. An Index of Sustainable Economic Welfare (ISEW) for Chile. *Ecological Economics*, Vol. 28: 231–244.

Crilly, M., A. Mannis, and K. Morrow 1999. Indicators for Change: Taking a Lead. *Local Environment*, Vol. 4, No. 2: 151–168.

Daly, H.E. and J. B. Cobb, Jr. 1989 (with contributions by C.W. Cobb). *For the Common Good: Redirecting the Economy toward Community, the Environment and a Sustainable Future*. Boston, MA: Beacon Press.

European Environment Agency (EEA) 2002. *Towards an Urban Atlas: Assessment of Spatial Data on 25 European Cities and Urban Areas*. Copenhagen.

Frazier, C. and K. Kockelman 2004. *Cities & Satellite Imagery: Models for Regional Change*. Paper submitted for presentation at the 2004 Annual Meeting of the Transportation Research Board, Washington, DC.

Glaeser, E. 1998. Are Cities Dying? *The Journal of Economic Perspectives*, Vol. 12, No. 2: 139–160.

Gustavson, K., S.C. Lonergan, and J. Ruitenbeek 1999. Selection and Modeling of Sustainable Development Indicators: a Case Study of the Fraser River Basin, British Columbia. *Ecological Economics*, Vol. 28: 117–132.

Hamilton, K. and M. Clemens 1999. Genuine Savings Rates in Developing Countries. *The World Bank Economic Review*, Vol. 13, No. 2: 333–356.

Hemphill, L., J. Berry, and S. McGreal 2004a. An Indicator-Based Approach to Measuring Sustainable Urban Regeneration Performance: Part 1, Conceptual Foundations and Methodological Framework. *Urban Studies*, Vol. 41, No. 4: 725–755.

Hemphill, L., S. McGreal, and J. Berry 2004b. An Indicator-Based Approach to Measuring Sustainable Urban Regeneration Performance: Part 2, Empirical Evaluation and Case-study Analysis. *Urban Studies*, Vol. 41, No. 4: 757–772.

Herrera-Ulloa, A. et al. 2003. A Regional-Scale Sustainable Development Index: the Case of Baja California Sur, Mexico. *International Journal of Sustainable Development and World Ecology*, Vol. 10, No. 4: 353–360.

Hodge, G. 1963. Use and Mis-Use of Measurement Scales in City Planning. *AIP Journal*, May: 121.

Hodge, R.A., P. Hardi, and D. Bell 1999. *Seeing Change through the Lens of Sustainability*. Background Paper for the Workshop "Beyond Delusion: Science and Policy Dialogue on Designing Effective Indicators of Sustainable Development," The International Institute For Sustainable Development, Costa Rica, May 6–9.

Innes, J.E. and D.E. Booher 1999. Metropolitan Development as a Complex System: A New Approach to Sustainability. *Economic Development Quarterly*, Vol. 13, No. 2: 141–156.

Kain, J.-H. 2003. *On Planning, Design, Decision-Making and Implementation of Measures for Sustainable Urban Development.* Chalmers University, Architecture, Working paper 030929.

Keiner, M. 2002. Indicator Based Control of Regional Planning. *Australian Planner,* Vol. 39, No. 4: 205–210.

Leaf, M. 2004. A Question of Boundaries: Planning and Asian Urban Transitions. In M. Keiner et al. (Eds.) *Sustainability and Urban Growth in Developing Countries.* Aldershot: Ashgate Publishers.

Leitman, J. 1999. Can City QOL Indicators be Objective and Relevant? Towards a Participatory Tool for Sustaining Urban Development. *Local Environment*, Vol. 4, No. 2: 169–180.

Levett, R. 1998. Footprinting: a Great Step Forward, but Tread Carefully – A Response to Mathis Wackernagel. *Local Environment*, Vol. 3, No. 1: 67–74.

Max-Neef, M.A. 1995. Economic Growth and Quality of Life: a Threshold Hypothesis. *Ecological Economics*, Vol. 15: 115–118.

Meier, K.J. and J.F. Brudney 2002. *Applied Statistics for Public Administration.* Fifth Edition. Belmont, CA: Wadsworth/Thomson.

Meyer, M. and E. Miller 2001. *Urban Transportation Planning: A Decision-Oriented Approach.* Second Edition. New York: McGraw-Hill.

MIT-CDD 2003. *Santiago Urban Planning Studio: Sustainable San Joaquín.* City Design and Development Group, Department of Urban Studies and Planning, Massachusetts Institute of Technology, Cambridge, MA.

Mitchell, G. 2000. Indicators as Tools to Guide Progress on the Sustainable Development Pathway. In J.R. Lawrence (Ed.) *Sustaining Human Settlement: A Challenge for the New Millennium.* North Shields: Urban International Press.

Newman, P. and J. Kenworthy 1999. *Sustainability and Cities: Overcoming Automobile Dependence.* Washington, DC: Island Press.

Neumayer, E. 2000. On the Methodology of ISEW, GPI and Related Measures: some Constructive Suggestions and some Doubt on the "Threshold" Hypothesis. *Ecological Economics*, Vol. 34: 347–361.

Neumayer, E. 2001. The Human Development Index and Sustainability – a Constructive Proposal. *Ecological Economics*, Vol. 39: 101–114.

Neumayer, E. 2003. *Sustainability and Well-Being Indicators.* Draft paper prepared for UNU-WIDER, July.

OECD 1978. *Urban Environmental Indicators.* Organization for Economic Cooperation and Development. Paris.

OECD 1997. *Better Understanding Our Cities: The Role of Urban Indicators.* Organization for Economic Cooperation and Development. Paris.

Pastille Project Consortium (PPC) 2002. *Indicators into Action: Local Sustainability Indicator Sets in Their Context.* London.

Pinfield, G. 1997. The Use of Indicators in Local Sustainable Development Planning: a Response to Jeb Brugmann. *Local Environment*, Vol. 2, No. 2: 185–187.

Portney, K.E. 2003. *Taking Sustainable Cities Seriously: Economic Development, the Environment, and Quality of Life in American Cities.* Cambridge, MA: MIT Press.

Quiroga, M.R. 2001. Indicadores de sostenibilidad ambiental y de desarrollo sostenible: estado del arte y Perspectivas. *Manuales No. 16*, UN CEPAL, División de Medio Ambiente y Asentamientos Humanos, Santiago de Chile, September.

Rees, W. 1997. Is "Sustainable City" an Oxymoron? *Local Environment*, Vol. 2, No. 3: 303–310.

Ryan, S. and J. Throgmorton 2003. Sustainable Transportation and Land Development on the Periphery: A Case Study of Freiburg, Germany and Chula Vista, California. *Transportation Research: Part D: Transport and Environment*, Vol. 8, No. 1: 37–52.

Segnestam, L. 1999. *Environmental Performance Indicators: A Second Edition Note*. World Bank Environment Department, Environmental Economics Series, Paper No. 71.

Selman, P. and Parker, J. 1997. Citizenship, Civicness and Social Capital in Local Agenda 21. *Local Environment*, Vol. 2, No. 2: 171–185.

Sistema Nacional de Información Ambiental (SiNIA) 2004. *Lista indicadores por Región. Región: Metropolitana.* http.//www.sinia.cl/indicadores/index.php?pag=region&r=RM, retrieved on April 22, 2004.

Sistema Nacional de Información Territorial (SNIT) 2003. *Plan Nacional de Captura y Estandarización de la Información Territorial 2003–2005: Estado de Avance Año 2003*. Santiago.

Smith, R. and R. Walker 1994. The Role of Performance Indicators in Housing Management: a Critique. *Environment and Planning A*, Vol. 26, No. 4: 609–621.

Spreng, D. and A. Wils 1996. *Indicators of Sustainability: Indicators in Various Scientific Disciplines.* Paper written in support of the project Multidimensional Approaches to Sustainability: The Framing Project proposed to the Alliance for Global Sustainability, September (revised October 2000).

Stirling, A. 1999. The Appraisal of Sustainability: some Problems and Possible Responses. *Local Environment*, Vol. 4, No. 2: 111–135.

Sussman, J. 1999. The New Transportation Faculty: The Evolution to Engineering Systems. *Transportation Quarterly*, Summer.

Sustainable Seattle 1998. *Indicators of Sustainable Community 1998*. Reprint 2004. Seattle.

United Nations Development Programme (UNDP) 2001. *Human Development Report 2001: Making New Technologies Work for Human Development*. New York and Oxford: Oxford University Press.

United Nations Division for Sustainable Development (UN DSD) 2004. http://www.un.org/esa/sustdev/natlinfo/indicators/isd.htm, retrieved April 9, 2004.

United Nations Human Settlements Programme (UN HABITAT) 1998. *Global Urban Indicators Database 2 (1998* data). http://www.unchs.org/programmes/guo/documents/1998.zip, retrieved on November 20, 2002.

United Nations Human Settlements Programme (UN HABITAT) 2004. *Global Urban Observatory*. http://www.unchs.org/programmes/guo/urban_indicators.asp, retrieved on May 6, 2004.

U.S. EPA 2002. *Smart Growth Index: A Sketch Tool for Community Planning Version 2.0.* Prepared for the U.S. Environmental Protection Agency by Criterion Planners/Engineers, Inc., November.

Wackernagel, M. 1998. The Ecological Footprint of Santiago de Chile. *Local Environment*, Vol. 3, No. 1: 5–25.

Wackernagel, M. and W. Rees 1996. *Our Ecological Footprint: Reducing Human Impact on the Earth*. Gabriola Island, BC Canada: New Society Publishers.

Weng, Z. and S. Yang. 2003. An Approach to Evaluation of Sustainability for Guangzhou's Urban Ecosystem. *International Journal of Sustainable Development and World Ecology*, Vol. 10, No. 1: 69–81.

Wheeler, S.M. 2000. Planning for Metropolitan Sustainability. *Journal of Planning Education and Research*, Vol. 20: 133–145.

World Bank 2003. *World Development Indicators 2003*. Washington, DC.

Chapter 7

TOWARD SUSTAINABLE URBAN MANAGEMENT IN DEVELOPING COUNTRIES

A Synthesis of the DIMSUD Project Findings and Further General Recommendations

Marco Keiner[1], Diego Salmerón[1], Willy A. Schmid[1], Iván Poduje[2]
[1]ETH Zurich, [2]Pontificia Universidad Católica de Chile

Abstract: This chapter concludes the findings of the completed research and positions the three case study cities with regard to their sustainability goals. One overall finding is that sustainability has to be defined for each specific context and that good urban management is the key to achieving it. In addition, recommendations for responsible urban management, such as establishing strategies for sustainable urban development, promoting decentralization, improving co-operation, harmonizing strategies with legal regulations and planning instruments, enhancing public participation and monitoring and controlling urban development, have been distilled. These recommendations, although based on the analyses of the case study cities, are also relevant for any other comparable city striving for sustainability. Finally, emphasizing the need to train national and local authority staff in the latest implementation methods developed from continuing research closes the gap between academia and practice.

Key words: Sustainable urban development, good governance, urban management, participation, planning tools, planning instruments, coordination, strategies, visions, monitoring, policy enforcement, training

M. Keiner et al. (eds.), From Understanding to Action, 191–217.

1. THE CHALLENGE FOR URBAN MANAGEMENT IN MEETING THE OBJECTIVES OF SUSTAINABLE URBAN DEVELOPMENT

1.1 Steps Toward Urban Sustainability

The global phenomenon of rapid urbanization is the main challenge to planning today. The convergence of economic growth, population dynamics and urban expansion offers both great challenges and great potentials for achieving urban-metropolitan sustainability. What can be done to ensure that urban sustainability becomes real and not only a utopian vision? To answer the main question of how sustainable urbanization can be achieved, Perlman et al. (1998) proposed the *Urban Sustainability Principles*:

1. There can be no global ecological sustainability without urban ecological sustainability.
   - Concentrating the human population in cities is essential to preserve both agricultural and wilderness areas.
   - Circular rather than linear urban systems are necessary to recapture resources.
2. There can be no urban environmental solution without alleviating urban poverty.
3. There can be no lasting solution to poverty or environmental degradation without a strong civil society and grassroots innovations.
   - Small may be beautiful, but it is still small.
   - Micro solutions need to be scaled up for macro impact.
4. There can be no impact of scale without
   - sharing what works among local leaders, or
   - scaling up into public policy.
5. There can be no urban transformation without
   - changing the old incentive systems and rules of the game;
   - forming collaborative partnerships among mutually distrustful sectors;
   - linking the local to the global through a transnational independent network.
6. There can be no sustainable city of the 21st century without social justice and political participation, as well as economic vitality and ecological regeneration.

These principles are interesting but somewhat dogmatic and difficult to concretize. What are needed are practical solutions. Achieving sustainability will require reduced resource use, improved performance of urban systems and processes, reduced waste and emissions and, ultimately, greater livability – ideas captured in the concept of the extended urban metabolism model (Asian Development Bank 2001, Newman and Kenworthy 1999). The urban metabolism model links together economic, social, environmental and cultural factors, providing a future-oriented perspective that recognizes the dynamics of the system and highlights the ultimate goals.

In order to give an overview of workable responses to urbanization problems, UN-HABITAT launched in 1997 the *Best Practices and Local Leadership Program* (BLP). Over 1600 proven solutions from more than 140 countries to the shared social, economic, and environmental problems of an urbanizing world are shown in an online database (UN-HABITAT 2002). Some of the best practices highlighted in the BLP database offer first steps towards achieving sustainability in Latin American and African cities. Achieving widespread implementation remains the greatest problem, because these "best practices" tend to take the form of specific principles for pushing forward sustainability within specific subsectors, such as housing, health care, and so forth. None of them actually represent "best practice" in a comprehensive sense. Others represent goals or objectives, but do not offer any suggestion of what sort of "practice" might actually help achieve these goals.

In the authors' experience, examples for good and best practices from developing countries can serve as models to effectively address chronic social, economic and environmental problems:

1. *General Urban Planning Practices*

- Renovation of public buildings, construction of points of identity (places, public buildings, parks, etc.)
- Comprehensive renewal and improvement of slum areas
- Definition of minimum quality standards for local road allowances, lot sizes and utility placement
- Compaction of urban development to prevent sprawl, for example, by brownfield redevelopment, by allowing higher densities for construction, by using unbuilt sites in the city before developing greenfields
- Enhancing energy-efficient settlement structures and housing
- Mixing functions and low and upper-income housing to overcome socio-economic segregation
- Controlling in-migration, not by limiting access to cities like in China or India, but by offering jobs in the countryside (for example, decentralization of labor intensive production sites in collaboration with national and regional planning authorities)

2. *Urban Infrastructure and Services Practices*

- Maintenance, renewal, and restoration of infrastructures (e.g. water supply, sewage network, roads, public buildings and green areas) through community-based neighborhood rehabilitation programs

- Managing demand for road space and providing quality public mass transport by the more efficient use of existing transport facilities, supported by measures like area licensing schemes, highly integrated land use and transportation planning

3. *Environmental Management Practices*

- Implementing integrated solid waste management programs aimed at separating at source, composting, recycling, and constructing and operating sanitary landfill sites
- Cleanup and garbage collection and separation projects (including income generating recycling for micro enterprises), for example, carried out by youth and volunteer groups
- Closing of open dump sites, installation of sanitary landfills
- Reducing air-polluting emissions by cars, industry and soft-coal burning households through the increased usage of unleaded gas, waste gas purification, energy saving and greenhouse gas-reducing cookers
- Elaboration of a comprehensive, participatory environmental master plan
- Protection of urban nature (biotopes, greenbelts, etc) and urban agriculture

4. *Housing Practices*

- Development of alternative housing norms to reduce construction costs
- Making housing affordable and accessible to disadvantaged groups and lower income families, for example, by providing decent self-help housing and infrastructure on suitable land to squatter cooperatives on a cost-recovery basis, using sustainable housing loan schemes and income generation activities
- Popularization of new building technologies, allowing cheaper construction, energy saving strategies, healthier conditions, and increased safety

5. *Urban Governance and Women Empowerment Practices*

- Establishing a broad-based steering committee for urban development comprising all groups of society
- Promoting neighborhood-based development committees in the consultative process and in preparing development plans
- Decentralization of administrative and financial powers, for example, by introducing participatory budgeting of municipal finances (e.g. Porto Alegre in Brazil)
- Public-private partnerships, partnerships between administrators, residents and NGOs, as well as international agencies and foreign central governments and sister cities
- Initiating and supporting poverty alleviation initiatives through small payback credits or loans for income generating activities (micro enterprises, arts and crafts, including traditional handicrafts, especially for the empowerment of women)

- Capacity building within the city administration, aimed at improving the performance of public services and implementing sustainable development; introducing a reward system for excellence in public services

6. *Social Services Practices*

- Crime prevention initiatives, based on social control in communities
- Reducing youth unemployment by promoting basic education, skills training and cultural activities
- Community-based preparatory and coaching centers and residential camps for children living in slums and squatter settlements to keep them in school
- Literacy training programs

Urban sustainability can only be reached if sustainable strategies are worked out and implemented by the cities and affected territories. Urban and rural sustainability strategies have to go hand in hand. Another attempt strongly backed by the international donor community, to improve living conditions in rural areas in order to reduce the movement of labor and brain drain to the cities, did not lead to the expected results. Through this experience, it was discovered that strengthening rural development may be a two-edged sword:

- Offering better education in rural settlements may lead to an education-driven migration to higher schools that are situated in cities. Once far from the farm, young people seldom return to the rural milieu.
- Mechanization in agriculture as well as political failures (like the expropriation of white farmers in Zimbabwe) result in rural unemployment, stimulating more rural-urban migration.
- Improving rural roads facilitates the access to urban areas.

For most of the big cities in the developing world, some of the best practices highlighted above offer first steps towards achieving sustainability. However, moving to widespread implementation remains the greatest challenge, because, these practices generally take the form of specific principles for pushing forward sustainability within each subsector. None of them actually represents "best practice" in a comprehensive sense. Others present goals or objectives, but do not offer any suggestion of new planning practices to help achieve them.

1.2 The Need for Good Governance

The main difficulty in achieving sustainable urban development is the lack of good governance (cf. Satterthwaite 1996, Gilbert 1998, Andam 2004; see also Chapter 4, Section 9 and Chapter 5, Section 5). Thus, this final chapter is dedicated to showing the most obvious examples of unsuccessful urban governance in the case study cities and to give workable recommendations for the planning practice.

A number of authors underline different approaches, including the value of local communities and the importance of public participation (Douglass and Friedmann 1997, Malbert 1998, Holston 1999), a grassroots approach (Douglass 1995, Abers 1997 and 2001) and the decentralization and democratization of planning decisions (Sandercock 2002). As enforcement is an important issue, evaluating the effectiveness of these policies in terms of achieving the stated goals is imperative. Once the objectives are clear and measures are defined, the planning process towards a more sustainable city must be steered. Most important is to control urban growth. For this, the use of sustainability indicators (Mitchell 2000, Keiner 2002), monitoring (El-Shakhs 1997) and controlling mechanisms (Keiner 2004; see Chapter 5, Section 7) is indispensable. The use of *Geographic Information Systems* (GIS) by the planning authorities is also a must (Cavric and Mosha 2001). Moreover, increased cooperation between cities and researchers in the field and among cities in the southern hemisphere in general (Atkinson 2002), could contribute to mutual learning, synergistic effects and the exchange of expertise.

Besides urban governance, another major challenge for urban development is to eradicate or at least to alleviate poverty. Possible solutions to this problem have been worked out in the UK ESCOR program (1998–2000), among others, for Johannesburg (Beall et al. 2000) and Santiago (Rodriguez et al. 2000).

Other challenges of sustainable urban development include the control of socioeconomic segregation and the harmonization of visions, strategies, and planning instruments. Perhaps the greatest challenge is dealing with the overlapping of functions in governing structures and overcoming the high degree of centralization in local decisions. These points will be highlighted in the following sections.

1.3 Establishing Visions and Strategies for Sustainable Urban Development

In general terms, the notion of sustainable urban development appears to the majority as a "politically correct" concept with good intentions but lacking concrete repercussions in the arena of urban actions. It seems that sustainable urban development has neither been established as an important concept in the design and implementation of public policies, nor in the reading of urban growth trends. The future task and main challenge for the management and planning of hyper-growing cities will be to make the turnaround from a rapidly growing urban area with exponentially growing problems to a sustainable developing city. Taking this into account, a first approach places a greater emphasis on the clearer definition of what sustainable urban development is in the specific context of each city concerned.

One important prerequisite is to create a common understanding of what sustainable urban development entails. Global and urban sustainability is hindered by high levels of material and energy consumption and waste generation, selective pressures on endangered ecosystems and new hazards arising from technologies designed to meet the demands of the wealthy (IIED 2001). In more affluent cities and those experiencing rapid economic growth, it is important to address larger scale environmental burdens, preferably before they become embedded in the infrastructure of the city and lifestyle of the people living there. The World Summit of 1992 was held based on this premise and the recognition that humanity had reached a turning point.

This context also inspired the vision *Joburg 2030* (City of Johannesburg 2002; see Chapter 5, Section 1.5). The document first recognizes the unique status of the city as well as the key threats to that status. It then maps a developmental pathway highly focused on facilitating economic growth driven by private sector investment. The City Council plays a significant role in ensuring an adequate physical infrastructure as well as maintaining socio-economic and cultural foundations such as controlling crime and enhancing the literacy and skills base. The focus of *Joburg 2030* and the development challenges at the grassroots level will have to be addressed in a global context. The quest for development that ensures the non-depletion and non-degradation of resources and the environment has become a major factor to be addressed by all development agencies including local authorities.

For Johannesburg, sustainable urban development can be defined as

> *"Development which enhances the city's global and regional attractiveness, ensures that the city meets its fair contribution to the global, regional and national commitments in sustainable development while at the same time ensuring a sensitive response to the specific needs and resource opportunities and constraints of its diverse regions and citizen groups."* (Irurah and Pile 2003)

In contrast, there does not exist a formal single vision of Santiago de Chile. Each ministry responsible for development issues in the capital creates its own vision. A similar phenomenon occurs within the real estate market and the 40 municipalities, which due to the absence of a metropolitan government, compete among themselves to attract public and private investment. However, with a view towards the celebration of the Bicentennial of the Republic of Chile (2000–2010), two planning efforts have been made:

the elaboration of proposals for an *Urban Reform* project and the formulation of a *Regional Development Strategy 2000–2006.*

In Chile, more than 80 professionals and representatives of diverse institutions, private as well as public, including real estate agent associations founded the *Work Group for Urban Reform* (GTRU). From 2000–2001, this group formalized its proposals in a document entitled "Foundations for Reform of the City and Territory" (BRCT), which highlighted the reference framework for starting decisive action plans:

- Urban development as a key role in the democratic system of the country
- Flexible policies for dynamic and diverse cities in constant change
- Urban politics affecting people through significant economic and social effects
- The paradigmatic change of the integration of the city with its surrounding territories
- The central role of the State in the development of and quality of life in our cities
- The role of the market and private initiative in urban development
- The importance of an adequate government structure to manage the city

As a result of the Work Group, ten dimensions were defined which encompass the broad urban and territorial themes of Chile and which, in turn define the operative areas for action under Urban Reform:

- Management and integral planning of the national territory
- Integrated planning of transportation and land use
- The quality of life in cities
- Public space, participation and citizenship
- The market as an urban and territorial regulator
- Financing the management and development of cities
- Cities as centers for productive activity
- Instruments and mechanisms of urban and territorial regulation
- Incorporation of the environmental dimension in urban development
- Institutionalization and management of Urban Reform

However, concrete results from the *Urban Reform* were few. In part, these were proposals to modify the Law and General Ordinance of Urbanism and Construction in order to modernize instruments and incorporate mechanisms of urban management that facilitate dealings with the private sector.

One of the five key themes of Chile's *Regional Development Strategy* is "Sustainable Territorial Environment: A Region with a Balanced Future." With this theme, the *Regional Development Strategy* seeks to promote sustainable economic and environmental growth in a way that leads to an effective improvement in quality of life, preventing the exclusion and segregation of inhabitants and territorial and environmental deterioration. Although both approaches, the *Regional Development Strategy* and the *Urban Reform* project, propose adequate schemes for managing the city in a more sustainable manner, they fall short by proposing structural changes without specifying the channels that will implement them. In practice, the intentions, which seek to integrate visions with a sluggish and unclear bureaucratic structure, remain only paperwork.

In the case of Gaborone, no proper vision for sustainable urban development exists (Keiner and Salmerón 2003). Although some "good governance" policies that create opportunities for sustainable development are already in use, they are not sufficient as a basis for sustainable urban development. Today, there is still a quest to define Gaborone's sustainable identity as a whole and for the city's constitutive parts. The concept of sustainable development was already introduced in Botswana in 1990 with the *National Conservation Strategy (NCS)*. One basic objective of the NCS is to ensure that

> *"Future generations have access to capital stocks of natural resources, at least similar to those presently available"* (Government of Botswana 1990),

a more or less condensed definition of sustainable development. Although the NCS is meant mainly for protected areas like national parks and game and forest reserves, its meaning could also be applied to urban development. One key sentence in the NCS that could be included in an improved urban development policy or strategy is,

> *"Achievement of sustainable development calls for comprehensive evaluation of environmental and economic implications before [sic] major new developments are undertaken."* (Government of Botswana 1990)

If this exigency had been taken into consideration in Gaborone's development, then perhaps several of the unsustainable developments outlined in Chapter 4 could have been avoided. However, in Botswana,

> *"Concerns with sustainable city development are not yet prominent in local planning concerns"* (Molebatsi 1996).

In other words, a lot of work remains to be done. If further initiatives are not undertaken, difficult development problems like unordered and resource consuming settlement patterns will take hold and remain.

Fortunately, there is a strong basis to do this work: *Vision 2016* (Presidential Task Group for a Long Term Vision for Botswana 1997; see Chapter 5, Section 1.5). This vision is multidimensional, encompassing the economic as well as the social, environmental, political, cultural, and spiritual aspects of the lives of Botswana's people. *Vision 2016* outlines Botswana's future aspirations and seeks to propel the country's socioeconomic and political development into that of a competitive and prosperous nation. Thus, it delivers the elaborated framework for a comprehensive and general strategy for sustainable urban development, which not just offers inspired definitions and principles, but also hints on how such an agenda could be translated into practice. Strong urban sustainability agendas place a similar emphasis on practical implementation steps.

1.4 Decentralization and Improved Multi-Level Cooperation

If sustainable development is to be achieved, appropriate measures have to be taken according to cities' assessed needs. In one city, the health care may be quite good due to high government investment in hospitals enhancing social sustainability. However, high levels of air polluting emissions from industry may simultaneously diminish environmental sustainability. In another city, the situation may be just the opposite.

Thus, the questions to be posed are *who* can determine what each city needs in order to become sustainable, and *what* is a priority and what is not? This centers the discussion on the legitimacy of the decision-making process and its subjects. Strictly speaking, in a democratic society, the citizens and their representatives are the ones who should be able to determine their needs and the direction of development.

This assumption has various implications:

- The necessity to understand sustainable development as a collective construct between legitimate actors in the direction it takes and in the determination of the requirements to attain its objectives.
- The necessity to accept that sustainable development does not have an a priori definition but is rather the result of specific relationships in territories with their own particular characteristics.
- The necessity to recognize sustainable development as a dynamic concept due to the fact that it is the object of social relationships situated in time and space. Consequently, its priorities and directions are subject to ongoing evaluation and transformation.
- The necessity to consider the diversity of sustainable development, that is, its diverse manifestations, orientations, contents and emphases.

The question of who decides what is needed and what has to be done varies from city to city and country to country. One main prerequisite is the vertical distribution of power between the national government and smaller territorial units, including the cities.

As Hall and Pfeiffer (2000) state,

> *"Successful urban strategies will be possible only if national and local governments work in close cooperation, if central governments define more clearly the most efficient distribution of functions between the dif-*

ferent levels of government, and if political activities follow a common framework."

In practice, many relevant decisions are made at the central government level in ministries that develop policies without a great deal of participation by local players.

For example, Gaborone's City Council (GCC) receives its public finances to a large extent from the Ministry of Local Government and the Ministry of Finance and Development Planning (see Chapter 5, Section 10.2). Therefore, the GCC is quite dependent on the central government to increase the revenues in order to compete with the rapid growth in expenditures. Since the GCC's expenditures far exceed its revenues, the revenue support grant from the central government covers at present nearly 70% of recurring expenditure. Development expenditures are funded by a 100% grant from the central government. Gaborone's City Council is also responsible for constructing and managing a large percentage of the city's social infrastructure. Although the GCC has the power to make bylaws for the city area, prescribing what licenses and permits are required and what fees must be paid for acquiring them,

> *"No bylaw made by a council shall be of any force and effect unless it is approved by the Minister and published in the Gazette."* (Townships Act; Government of Botswana 1999)

Any changes proposed in these fees are subject to another bylaw, which similarly requires the Minister's approval.

The city of Gaborone does not have the necessary power to decide how public finances should be spent on sustainability-oriented projects. The requirement of approval from superior authorities may cause delays in the implementation of the city's measures and plans.

In conclusion, decentralization and the transfer of power and decision-making capability to the municipal level is a must for sustainable urban development. However, despite attempts to decentralize functions in many developing countries, the autonomy and power of regional and local governments continue to remain fairly modest.

1.5 Improved Horizontal Cooperation

Another point for consideration is the *horizontal overlapping of functions* and power of authorities:

> *"[...] Government structures need to work horizontally in order to implement the interdisciplinary nature of sustainable development. Governance is of core importance in implementing decisions towards sus-*

tainability and in effectively managing public interests in a politically organized community." (FARN, quoted in The Regional Environmental Center 2004)

One example for unsuccessful horizontal coordination can be seen in Santiago de Chile. In Santiago, public transport is provided through about 9,000 city buses running on 354 lines. The large number of buses overloads the main roads and results in long cross-city rides, which can take over an hour and a half in travel time (Malbrán 1998). There are also three metro lines covering a total of 40 kilometers, plus taxis. These means of transportation do not operate as an integrated network, causing insufficient quality of service and a reduction in public transport efficiency. An effort to revert this situation has been made by means of the Santiago Urban Transportation Plan, known as *Transantiago*. In addition, a number of public institutions are attempting to address issues related to urban development, but rarely in a coordinated fashion. There are at least four institutions responsible for infrastructure and transportation at the central level: the *Ministry of Public Works* (MOP), the *Ministry of Housing and Urbanism* (MINVU) operating through the *Housing and Urbanism Service* (SERVIU), the *Metro*, and the *Ministry of Transportation and Telecommunications* (MTT). The MOP is in charge of planning and building interurban roads (major highways) as well as public buildings like airports, prisons, hospitals, and schools. With Public Works Concessions, the MOP has been able to intervene more directly in the city, even though the planning and construction of roads on this scale is under the jurisdiction of MINVU. Since this organization's limited resources are mostly allocated to public housing, many main highways are transferred by executive decree to the MOP, further complicating management. The MTT, on the other hand, is officially in charge of public transportation. The same minister of the MOP directs the MTT, however this does not guarantee coordination because the departments continue to duplicate functions in many cases. Thus, the implementation of *Transantiago* has been slower than expected, and many of the works scheduled for 2005, when the current political administration ends, will probably not be implemented.

No single institution has the formal task of analyzing urban expansion in Santiago and its effects on sustainability to cope with the problem of the city's growth. As described above, this responsibility is distributed among different public agencies and some non-governmental organizations.

These examples show that urban policy and planning requires a clear distribution of functions and power. Obviously, the basic lessons from "New Public Management" have not yet been sufficiently applied.

A third aspect worth mentioning is the coordination between public institutions and the private sector. It must be recognized that the dynamics of development are constantly exposed to pressures from industry, the economy and above all in the case of Santiago, from the real estate market. This fact is a problem related to the modification or adaptability of master plans according to market pressures. Since the definition of zoning regulations depends on subjective criteria, the planners are able to modify the regulations at will, arbitrarily generating a great deal of wealth for particular parcels on which potential investment is focused. On the other hand, a local zoning plan may define building conditions that are never implemented, because the market does not end up occupying those spaces, as happens in poor neighborhoods.

As one can see in this quick summary, Chile urgently needs to integrate responsibilities and powers to insure the proper coordination of planning measures in the capital city. A broad law is needed that coordinates the various different norms and unifies basic criteria on zoning and investments.

The last point to mention in this section refers to the will or ability to implement plans. Again, an example from Santiago will be stressed. The *Plan for the Prevention and Atmospheric Decontamination* (PPDA) of the Metropolitan Region of Santiago was formulated in 1997. Although the distribution of power is clear, as the *National Commission on the Environment* (CONAMA) has assumed the leadership in the design and execution of the PPDA, difficulties in reaching agreement have arisen, complicating the management of the plan. Along the same lines, another analysis views the various institutions as the biggest problem facing the implementation of the plan:

> *"In the application of the PPDA, it is evident that the institutions and a large part of society are not prepared to take on the plan and are not the least bit interested in making favorable interventions in the part of the system which corresponds to them."* (Rodrigo Jiliberto, quoted in de la Cuadra 2000)

This analysis coincides with the conclusions of the study by Nicod and Iizuka (2000) on citizens' perception of environmental pollution.

1.6 Harmonizing Strategies, Planning Instruments, and Legal Regulations

Urban policies are applied through a set of planning instruments and bylaws that range from the regulation of land use to the definition of public and private transportation systems. Requirements for development to proceed as outlined in a city's visions and strategies include the following:

- Plans must be oriented toward the objectives of the conceptual framework (i.e. focused on sustainable development)
- Plans and bylaws on urban development should not contradict other plans, strategies, and legal regulations

– Plans should cover an appropriate spatial perimeter, i.e. take the urban-rural link into account

Most of the assessed policies and strategies relating to urban development do not have a clear hierarchy of goals and objectives, and the link between goals and their measures is usually missing (see Chapter 5, Section 5.1). In addition, measures sometimes get mixed with goals, which complicates the clarity of a plan or policy. To improve the performance of the different plans, a clear hierarchy of goals is necessary.

In Chile, despite efforts in formulating a nationally coherent urban policy, no such policy has been formally approved. This lack of national guidelines acts as a repercussive vacuum, where the nation's regions and communities still lack a clear planning framework. Thus, planning in Santiago can be explained as the sum of uncoordinated legal regulations and sector plans, which often contradict each other. For example, although the *Metropolitan Santiago Master Plan* (PRMS; see Chapter 5 Section 3.3) defines a boundary beyond which no kind of urban occupation should occur unless it is intended for agriculture, agribusiness activities, or the protection of the natural environment, a specific law, *Law DL 3156 on Agricultural Land*, permits the subdivision of any land into lots of up to 5,000 m^2 through approval by the *Agricultural and Livestock Service* (SAG) of the *Ministry of Agriculture*. Authorization by the *Ministry of Housing and Urban Planning* (MINVU) or the municipalities is not required when the land use change to non-residential uses is not in question. Used as a legal loophole, *Law 3156* permitted the registration of more than 30,000 hectares for use as rural residential properties, before the urban extension boundary was approved. This situation was possible because of an overlapping of functions, giving the *Agricultural and Livestock Service* (SAG) jurisdiction on this matter, which clearly should be under the sole jurisdiction of metropolitan planning. Moreover, in 1997, the PRMS incorporated an additional 12,000 hectares north of Santiago. On the one hand, this extension drew strong criticism since it openly contradicted the policy of city densification by duplicating Santiago's expansive growth pattern. However, in relation to this expansion, new laws were established to allow development to become subject to certain conditions, authorizing real estate development provided the developer not only assumes responsibility for all services but also pays for impacts on transportation and on the environment (a form of impact/pollution tax or levy). These conditions open up a new concept in the urban regulation of Santiago. For the first time, environmental and transportation impacts were estimated before subdivisions were approved, fixing mitigation requirements on a metropolitan scale. In addition, measures were taken to charge the developers significant amounts of money to finance related infrastructure projects, including treatment plants and new roads.

Another example is *Decree 420* of 1979, which concerns the liberalization of urban land by creating an *Urban Expansion Area*. This decree also derogates the objectives of the PRMS, dedicated to the location of public housing and equipment, and it removes the restrictive conditions on growth and extension. Although MINVU reestablished some explicit regulations to prevent the unlimited expansion of the city in 1985, it is still possible to carry out unforeseen real estate development under certain conditions. Public agents, mainly SERVIU, which builds public housing, exercise pressure in this direction to allow private real estate promoters to obtain cheap, new urban land. This practice openly contradicts the policy of urban densification, encouraging Santiago's expansive growth trend.

To cope with these problems, the project *Ordenamiento Territorial Región Metropolitana Santiago de Chile* (OTAS) seeks to implement a *Regional Policy of Territorial Zoning* to prepare the technical foundations to orient and define a policy of territorial zoning. This will lead to environmentally sustainable development in the metropolitan region by proposing instruments and designing formal procedures for institutional coordination and citizen participation at the regional level, which in turn can serve as a pilot program applicable to other regions or territorial levels. For this, the regional government of the metropolitan region has to be strengthened as manager of environmentally sustainable territorial zoning. There are many more examples that show Chile's urgent need to integrate responsibilities and powers regarding the effective management of its territory.

In Johannesburg, in spite of the strong alignment between the municipal *Integrated Development Plans* (IDP) and *Local Agenda 21* principles and processes, there has not been an official recognition of the opportunity to apply sustainable urban development as the overall framework for the IDP process and its outcomes (see Chapter 5, Section 6.2). Consequently, urban sustainability strategies have not been explicitly applied in the conceptualization of the IDPs. Also, there is no systematic mechanism for linking the monitoring and reporting of outcomes (from the *State of the Environment Report*, City of Johannesburg 2003, for example) to the other sector plans to ensure action aimed at improving indicators in an integrated and synergetic approach. In other words, there has not been an explicit effort towards closing the loop of sustainable urban development (planning, implementation, monitoring, evaluation, and a return to the planning stage). Recasting *Joburg 2030* and future cycles of IDPs (including related *Spatial Development Frameworks* and other sector plans) with the broader framework of urban sustainability would be a crucial step in the right direction. This could be followed by an expansion of the Key Performance Areas and IDP related indicators of the city's performance scorecard to include operational areas and management as well as broader urban sustainability indicators (see

Chapter 6). These indicators facilitate and support monitoring and evaluation for both categories of issues rather than the management indicators only. Johannesburg must refocus its attention on this need in order to sustain its global and regional competitiveness and also ensure that grassroots needs will be met in a sustainable manner over time.

In Botswana, due to the fact that modern physical planning in the country is still a relatively new topic, there are also inadequacies concerning the co-ordination of plans and policies and the organization of planning procedures. The present statutory framework regulating town and country planning in its present form is inadequate and out of line with current environmental policy as stated in the *National Conservation Strategy* of 1990. In general, environmental law in Botswana is fragmented in various different pieces of legislation administered through a number of different administrative bodies, resulting in incoherent decision-making. There are also some major problems concerning the coordination between plans and policies as well as the lack of their systematic monitoring and review. Approval of plans is slow. The *Gaborone City Development Plan* (GCDP), for example, refers to the time period from 1997 to 2021. It was approved in 2001 at the time when the first review should have already been made. Due to rapid growth and change of the built environment of Gaborone and the surrounding areas, some of the assumptions made in the plan no longer reflected reality at the time of approval. The *Greater Gaborone Structure Plan* (GGSP), which should officially determine general land uses has never been approved, the *Gaborone Central Business District Master Plan* of 1994 was delayed by politics until now and the *Gaborone City Landscaping Master Plan* of 1995 has no legal backing, but is merely advisory for planners and developers.

Given the many plans that are not in use or are already outdated when they are enacted, the DIMSUD project recommends a major review of Botswana's planning system. First, the envisaged and long debated *National Physical Plan* (NPP), which provides a spatial framework for the coordination and implementation of development programs and projects at the national, regional, and community level, should be completed and approved. In addition, the *South East District Settlement Strategy* (SEDSS) and the *South East Region Master Plan* (SERM), which are in preparation but not yet implemented, should be used to guide Gaborone's physical development planning. In addition, the goals and objectives of the *Vision 2016*, adjusted to the needs of Gaborone, should be included in the SEDSS.

To adapt the planning system to the realities of spatial development, it must be adjusted to the right perimeter. Thus, the differentiation between "tribal land" and declared "planning areas" should be abolished. The entire urban area of Gaborone should be determined a "planning area" and the *Town and Country Planning Act*'s declaration of the City Council as the planning authority in charge, put into practice. According to the *Tribal Land Act*, Land Boards are in charge of the land use zoning and planning of tribal land, i.e. the many villages surrounding Gaborone. Planning laws only apply to planning areas, thus, the development of tribal land, which lies de facto in the urban agglomeration of the capital, is not included. So, the new *South East District Settlement Strategy* should merge the tribal lands and planning areas of Gaborone by including the entire area of the City of Gaborone, even if it does not belong all to the South East District. In actuality, this district surrounds the administratively isolated capital. The *Greater Gaborone Structure Plan* (GGSP) should subsequently be fostered and amended to become a detailed "Coordination Plan." Due to the findings of *the South East District Settlement Strategy*, detailed alternatives to development should be elaborated. Finally, the economy-focused *Gaborone City Development Plan* (GCDP) should be subordinated to the physical GGSP.

Sustainability-oriented planning in Gaborone could be enhanced with the creation of a *Committee for Sustainable Urban Developmen*t (CSUD), which should function under the Physical Planning Unit of Gaborone's City Council. Its task would be to provide the local and national authorities with a recurrent *Report on Sustainable Urban Development* (RSUD). This report should contain the results of indicator analysis and recommendations to political decision-makers and planning authorities for further development towards sustainability.

1.7 Public Participation

As shown throughout this book, public participation on a grassroots level is an indispensable part of sustainable urban development. Any measures aimed at achieving sustainable urban development in Africa and Latin America imply prerequisites on the national level, such as decentralization, democratization, "good governance," encouragement of community participation and integration of women, family planning and health campaigns to cope with the problem of HIV/AIDS, an appropriate education system as well as the allocation of credit and capital for micro enterprises and self-help associations.

FARN (quoted in The Regional Environmental Center 2004) reinforces the importance of public participation:

> *"Public participation, different from the one existing in traditional representative democracies, must be present. Thus, a sustainable city requires institutions and systems that can facilitate public participation in decision-making regarding environmental use and management."*

In the current plans and policies of the case study cities, good as well as bad examples of public participation can be found.

In Santiago, little transparency in public decisions on urban regulation, which tend to favor private real estate investment, is apparent. During the development of the program *Sustainable Chile*, social leaders of the Metropolitan Region of Santiago and local community representatives asked for better public participation and criticized the existence of an uninformed citizenry and deteriorated social relationships, a lack of confidence in the government and individualism.

In Botswana, the government has recognized the importance of involving its people in the planning process. Planning participation and consultation structures in Botswana, some of which are officially defined, exist at various levels of society. Some consist of elected representatives while others are formed on an ad hoc or voluntary basis. Formal structures like Village Development Committees, Council Committees and the City Council itself, Land Boards, Tribal Administration, District Administration and Parliament are already taking part in the planning process. In the following paragraph, the model of *Kgotla* (see Chapter 3, Section 3.4) is highlighted.

The Kgotla is a traditional instrument for articulating public and individual interests in Botswana. Its functions are similar to those of a local parliament. The Kgotla is a meeting place in a ward. Wards are subsets of Council political areas, and there are 26 wards in Gaborone. The Kgotla is regarded as a legitimate institution for public decision-making. Once a decision has been made at the Kgotla, it is considered binding for the entire community. The Kgotla holds public meetings for development activities concerning the city, in addition to its customary court functions. Kgotla forms an important forum that brings urban development issues to the attention of the government. It is the starting point for bottom up development and forms a basis for involving the people in decision-making processes. The execution arm of the Kgotla is the *Ward Development Committee (WDC)*. It represents the lowest level in the hierarchy of planning bodies. Apart from ex-officio members (councilors, officers of the city council), the members of the WDC are elected by the Kgotla and are responsible for the following:

- Identifying the development needs of the community
- Coordinating development planning and projecting implementation within their area of jurisdiction
- Coordinating development activities of volunteer organizations
- Informing and assisting members of the public in the preparation of urban development plans, physical development plans and other policies and plans related to social issues

The importance of including the public cannot be stressed enough. Planning mistakes can be avoided and spatial conflicts can be minimized if the public is involved in all stages of plan preparation. Emphasis should be given to the consultation phase at the beginning of the planning process to further integrate the public. This early consultation also contributes to effective planning since plans do not always have to be adjusted radically in the proceeding planning period and a broad common sense for how development should proceed is established at an early stage.

In addition, the implementation of urban development plans and physical plans should ideally be initiated and backed by the authorities and taken up by community associations and NGOs. When public authorities are overextended, self-help communities, NGOs, and grassroots initiatives become more and more important. These organizations contribute to problem solving strategies on lower, basic levels, but lack access to political power and capital. Authorities should support such organizations instead of arguing that they should follow "official" policy.

Several possibilities to improve public participation exist. Basically, a distinction exists between *active* and *reactive* public participation. Active public participation begins at the first stages of the development of a physical plan, for example with the *Ward Development Committees* in Gaborone. The WDCs could also function as a source of information for the development of physical plans or policies. New interest groups or associations could also take part in round table discussions together with already existing NGOs. Performing workshops with experts on different subjects is another possibility for actively incorporating public needs into physical plans and policies. Nowadays, the opportunities the Internet offers should not be disregarded. Botswana's youth are very familiar with the Internet. Using the Internet to gather the needs, ideas, visions, and opinions of the public to conceive the overall guidelines for spatial development is worth a try.

In collaboration with the *Greater Johannesburg Metropolitan Council* (GJMC), the *International Council for Local Environmental Initiatives* (ICLEI) backed the *Land Development Objective* (LDO) since 1993, which proposes a new urban management process that aligns stakeholder-defined priorities with local government planning, budgets and personnel. To accomplish integrated environmental management and development in Greater Johannesburg, the LDO was directly aligned to the *Local Agenda 21* process. The LDO consists of seven sectors, one of which is environmental management. The *Environmental Management LDO* was formulated under the rubric of the *Development Facilitation Act* (DFA-1995), an act that focuses on achieving sustainable development from a financial and environmental perspective. The aim of the *Environmental Management LDO* is to achieve recognition for environmental management in local governments and to develop institutional structures and obtain resources to make this management effective. A partnership group composed of members from the public and private sectors and interest groups coordinated the overall planning process. In 1996, an Environmental Management Development Forum was established to identify and set priorities for the metropolitan area. Membership was open to all. The creation of this stakeholder group and community consultation were simultaneous and complementary processes. This participatory approach, including workshops, succeeded in translating environmental issues into a strategic focus of the GJMC. As an outcome, environmental management has been integrated into the institutional framework, and priorities (inclusive budgetary allocations) for action plan development have been set (ICLEI 2004b).

1.8 Monitoring and Controlling Sustainable Urban Development

In Santiago, mechanisms and impact control systems have been developed which seek to measure the environmental externalities of real estate development until now, passing these costs on to mitigation works. The previously mentioned project *Ordenamiento Territorial Región Metropolitana Santiago de Chile* (OTAS) produces a GIS platform which provides diagnostic coverage and plans of urban and territorial impact that serve to support policy and to orient sector lines of ministries and municipalities. However, this project still lies within academic circles and has very little influence on the actions of the ministries.

In Johannesburg, the *State of the Environment Report* (City of Johannesburg 2003) actually sets a strong base for status quo and future monitoring of key sustainability indicators (environmental, socioeconomic, cultural, and institutional). The document applies the DPSIR model (driving force, pressure, state, impacts, and responses) of data capture and reporting. With regular capture and reporting of such data, one can better track sustainability trends in the city.

Although indicator sets are a known tool in Botswana (for example, UNCHS indicators for the comparison of Gaborone to Francistown), a comprehensive indicator set for sustainable urban development on the city level for Gaborone does not yet exist.

As outlined in Chapter 6, Section 3.8, significant amount of data exists for the case study cities today, which could be used in sustainability oriented monitoring and controlling. At the same time, there is a lack of specific data in the area of spatial observation, and the management of all the available data is more or less uncoordinated. The availability and quality of the data therefore do not provide a comprehensive monitoring of sustainable spatial development and must be improved.

Since spatial development is a dynamic system, development planning must be a continuous process as well. The goals, measures, and instruments should be adapted continuously to the changing framework. As spatial development is very complex, it is important to use tools and models that reduce this complexity. The urban developments of the case study cities reveal that the systematic spatial observation and the definition of development goals are necessary in order to achieve sustainable urban development. Thus, monitoring and controlling concepts have been proposed by the DIMSUD project as tools to support the ongoing planning process and provide a comprehensive framework used for strategic and operational urban planning. This means collecting data and information in a systematic manner so that it can then be used for objective and well-founded decision-making.

Monitoring involves continuous spatial observation based on selected indicators, gathering information (for example, statistical or GIS analyses) covering a long time period and a wide thematic range. Monitoring is a systematic ongoing analysis of the environment and can reveal critical developments. Since monitoring always provides up-to-date information, development trends can be determined at an early stage. Indicators should cover a wide variety of topics and can also be used for controlling and evaluating urban development.

Controlling is a part of the monitoring process and compares the goals of the development plan with the current state of development (Keiner 2002). This comparison helps determine whether the goals and objectives have been or are being achieved and where measures must be taken in order to achieve them. Thus, basic information for the revision of strategies and plans can be provided. Whereas monitoring delivers extensive information about the current state of the environment, controlling only considers what is relevant with respect to the goals defined in the development plan and their associated processes. Monitoring and controlling can be combined to assess and evaluate spatial development and to gather information for further steps.

In general, one can conclude that the existence of a vision and an appropriate implementation structure still does not guarantee the successful integration of new initiatives. The involved parties must be committed to these goals and be able to work together to realize them.

2. LESSONS LEARNED AND OUTLOOK

Main conclusions of the DIMSUD project and of this book are:

- Sustainable urban development challenges vary among the cities studied.
- Given current development patterns each of the cities studied is not on a sustainable pathway.

- Despite these patterns there are opportunities for the case study cities to enhance their ability to achieve sustainable development.
- Research in urban sustainable development through cooperative links between institutions in both developed and developing countries provided for a valuable and fruitful experience.

Working with city-specific definitions of sustainable urban development will allow the monitoring and controlling of the development over time. For each city in a developing country striving for sustainable development, bundles of harmonized and appropriate measures must be kept together to maintain their effectiveness. This is of the utmost importance, if we consider that sustainable urban development assumes a balance between the needs of the population and the presence of specific resources, including the location of health and education services, recreation areas, adequate public transport, and citizens' security, among others.

Most developing regions are still going through urban transition and will retain a significant rural population for many decades. This means that sustainable development programs must reinforce urban-rural linkages. Sustainable development therefore calls for an ecosystem approach with the following elements:

- Inclusion of people and their activities in the ecosystem
- Awareness of ecosystem structure and function at multiple scales
- The use of ecological boundaries to define environmental planning, assessment and management of units
- Adaptive management strategies, based on feedback from new information to improve management and policies when they prove to be unstable
- Participatory management involving all stakeholder groups
- Integration of science and human values in the formulation of goals for protecting the ecosystem
- Recognition of the limits to action in ecosystems
- Geographically comprehensive systems with levels of analyses including interactions among physical, chemical, biological, and social components

National and local authorities should be aware of these requirements. For implementing policy reforms, governments need a strong and sustained political will. Also, they should be open to introducing innovative management systems and integrating environment and ecological issues into decision-making at all levels, giving them equal priority with economic and social concerns (Andam 2004). Thus, the development of human resources and institution building is indispensable.

DIMSUD is also a research project that offers a link to a training component. Each case city may apply for training carried out by The *International Council for Local Environmental Initiatives* (ICLEI). ICLEI is an international association of local governments implementing sustainable development. Its mission is to build and serve a worldwide movement of local governments to achieve tangible improvements in global sustainability with a special focus on environmental conditions through cumulative local actions. Through its campaigns, ICLEI helps local governments generate political awareness of key issues, build capacity through technical assistance and training and evaluate local and cumulative progress toward sustainable development. ICLEI serves as an information clearinghouse on sustainable development by providing policy guidance, training and technical assistance, and consultant services to increase the capacity of local governments to address global challenges (ICLEI 2004a).

Despite the lessons learned, experiences documented and recommendations given, what is the chance that sustainable development will really be effectively implemented? Sustainable Cities (1992) determined that most citizens of urban Africa, Asia, and Latin America would find it difficult to share the global environmental concerns of the North:

> *"Questions of survival 20 or more years into the future have little relevance to those concerned with survival today."* (Sustainable Cities 1992)

Cities need to be sustainable at all levels. Their internal environments must be clean and of high quality for their residents. At the same time, cities must close the loop in terms of interactions with the hinterland. That is, they must decrease metro-wide pollution levels. Ultimately, urban activities must be looked at in terms of their impact on the global environment, including the consumption of resources and the production of gasses that quickly affect the global scale. That is the challenge of creating sustainable cities. Urban design issues cut across all of these levels. Sustainable Cities (1992) doubt that the environment-driven approach towards sustainability, as it appears in Europe and North America through *Local Agenda 21* processes, would work for developing countries:

> *"It may be misleading to refer to many of the most pressing environmental problems in Third World cities as 'environmental' since they arise not from some particular shortage of an environmental resource but from economic or political factors which prevent poorer groups from obtaining them and from organizing to demand them."* (Sustainable Cities 1992)

Moreover, Rydin et al. (2003) doubt that sustainable development would be the most easily enacted response despite the obvious challenges of urbanization, because there lacks a consensus as to the meaning of sustainable development and quality of life, due to a real tension regarding future visions and what they entail. Also, Andam (2004) points out that there is a lack of vision and awareness in general, which has to do with illiteracy and language barriers in communication. For the example of Africa,

> *"Environment and development conventions are conducted in languages that most local people cannot fully understand."* (Andam 2004)

Therefore, it is indispensable that the concept of sustainability be translated into simple language and understandable implementations on the grassroots level and be based on a broad societal agreement that is fully assimilated into the culture of the organizations responsible for the implementation of sustainable urban development (Innes and Booher 2000).

Possible urban management policies are discussed in the literature, ranging from establishing innovation strategies and new technology (Brennan 1994, Sassen 1998), improving land tenure security (Gilbert 1994, de Soto 2000) and solutions for urban transportation (Rimmer 1998, Gakenheimer 1994), to measures to protect the climate (Liddle and Moavenzadeh 2002, Molina and Molina 2002). Von Weizsäcker et al. (1997), de Roo and Miller (2000), and Jenks and Burgess (2000) suggest that the structure of the built city should be as compact as possible for the efficient use of resources. This also would save financial resources that could be used to fight urban poverty and improve environmental conditions in the cities. The challenge is now to transform these scientific approaches into workable solutions in practice, solutions that vary from context to context, because there are no consumer-ready universally valid "recipes" for finding the path to sustainability

In conclusion, solutions for sustainable urban development have to meet specific local needs and cannot be applied in a copy-paste manner. Thus, for every city striving for sustainable development, specialized analyses must be made and an appropriate bundle of necessary measures to be taken by national, regional and local governments gathered together. In this sense, the DIMSUD project can offer specific solutions to specific problems in each of the case study cities. Other cities worldwide may be inspired by DIMSUD's methodology and findings and encouraged to try to find their own way to a sustainable future.

REFERENCES

Abers, R. 1997. Learning Democratic Practice: Distributing Government Resources through Popular Participation in Porto Alegre, Brazil. In M. Douglass and J. Friedmann (Eds.) *Cities for Citizens: Planning and the Rise of Civil Society in a Global Age.* Chichester (UK): John Wiley & Sons: 39–65.

Abers, R. 2001. Practicing Radical Democracy – Lessons from Brazil. *DISP* 147: 32–38.

Andam, K. 2004. *Main Challenges of Sustainable Urban Development in Fast Growing Cities of the South.* Keynote speech. Annual Meeting of the Alliance for Global Sustainability, Gothenburg, March 23, 2004.

Asian Development Bank 2001. *Urban Indicators for Managing Cities*. Manila.

Atkinson, A. 2002. International Co-operation in Pursuit of Sustainable Cities. In D. Westendorff and D. Eade (Eds.) *Development and Cities*. Oxford: Oxfam.

Beall, J., O. Crankshaw, and S. Parnell 2000. *Johannesburg*. Working Paper 12, Urban Governance, Partnership and Poverty. Birmingham: International Development Department, School of Public Policy, The University of Birmingham.

Brennan, E. 1994. Mega City Management and Innovation Strategies – Regional Views. In R.J. Fuchs, E. Brennan et al. (Eds.) *Mega City Growth and the Future*. Tokyo, New York, Paris: UNU Press: 233–255.

Cavric, B. and A.C. Mosha 2001. *Towards Better Urban Development and Management in Botswana*. Paper presented at the AESOP Conference at Shanghai, July 11–15, 2001.

City of Johannesburg 2002. *Joburg 2030 – Full Report*. http://www.joburg.org.za/joburg_2030/2030_documents.stm, retrieved on March 2, 2004.

City of Johannesburg 2003. *State of Environment Report 2003*. Johannesburg.

De la Cuadra, F. M. 2000. *Conflicto ambiental y acción colectiva en Chile: el caso de la contaminación atmosférica de Santiago*. http://168.96.200.17/ar/libros/ecologia/cuadra.pdf, retrieved on March 12, 2004.

Douglass, M. 1995. *Urban Environmental Management at the Grass Roots. Toward a Theory of Community Activation*. East-West Center Working Papers, No. 42. Honolulu.

Douglass, M. and J. Friedmann (Eds.) 1997. *Cities for Citizens: Planning and the Rise of Civil Society in a Global Age*. Chichester (UK): John Wiley & Sons.

El-Shakhs, S. 1997. Towards Appropriate Urban Development Policy in Emerging Mega Cities in Africa. In C. Rakodi (Ed.) *The Urban Challenge in Africa – Growth and Management of its Large Cities*. Tokyo, New York, Paris: UNU Press: 497–526.

Gakenheimer, R. 1994. Six Strategic Decision for Transportation in Mega Cities. In R.J. Fuchs, E. Brennan et al. (Eds.) *Mega City Growth and the Future*. Tokyo, New York, Paris: UNU Press: 322–348.

Gilbert, A. 1994. Land Shelter in Mega Cities: Some Critical Issues. In R.J. Fuchs, E. Brennan et al. (Eds.) *Mega City Growth and the Future*. Tokyo, New York, Paris: UNU Press.

Gilbert, A. 1998. World Cities and the Urban Future: The View from Latin America. In F.-C. Lo and Y.-M. Yeung (Eds.) *Globalization and the World of Large Cities*. Tokyo, New York, Paris: UNU Press: 174–202.

Government of Botswana 1990. National Conservation Strategy. *Government Paper No. 1 of 1990*. Gaborone: Government Printer.

Government of Botswana 1999. *Townships Act*. Gaborone: Government Printer.

Hall, P. and U. Pfeiffer 2000. *Urban Future 21: A Global Agenda for Twenty-First Century Cities*. New York: Spon Press.

Holston, J. 1999. *Cities and Citizenship*. Durham: Duke University Press.

ICLEI 2004a. *About ICLEI*. http://www.iclei.org/about.htm, retrieved on March 4, 2004.

ICLEI 2004b. *Johannesburg, South Africa*. http://www.iclei.org/LA21/cities/Johannesburg.htm, retrieved on April 19, 2004.

IIED 2001. *Economic Growth and Shifting Urban Environmental Burdens*. http://www.iied.org /docs/urban/urbenvir_bp03.pdf, retrieved on March 3, 2004.

Innes, J. and D. Booher 2000. Indicators for Sustainable Communities: a Strategy Building on Complexity Theory and Distributed Intelligence. *Planning Theory and Practice* Vol. 12, No. 1: 173–186.

Irurah, D. and K. Pile 2003. DIMSUD Project: Johannesburg. Final report. Johannesburg: University of the Witwatersrand.

Jenks, M. and R. Burgess (Eds.) 2000. *Compact Cities – Sustainable Urban Forms for Developing Countries*. London: Spon Press.

Keiner, M. 2002. Indicator Based Control of Regional Planning. *Australian Planner* Vol. 39, No. 4: 205–210.

Keiner, M. 2004. Sustainable Development and Urban Management in Developing Countries: The Case of Africa. In M. Koll-Schretzenmayr, M. Keiner, and G. Nussbaumer (Eds.): *The Real and the Virtual Worlds of Spatial Planning*. Springer: Berlin, Heidelberg: 43–59.

Keiner, M. and D. Salmerón 2003. Sustainable Urban Development in Gaborone, Botswana. DIMSUD working paper. *ETH e-Collection*. http://e-collection.ethbib.ethz.ch/cgi-bin/show.pl?type=bericht&nr=308, retrieved on May 6, 2004.

Liddle, B. and F. Moavenzadeh 2002. The Sustainability Challenge for Climate Change: Balancing Inter- and Intergenerational Equity. In F. Moavenzadeh, K. Hanaki and P. Baccini (Eds.) *Future Cities: Dynamics and Sustainability*. Alliance for Global Sustainability Bookseries Science and Technology: Tools for Sustainable Development, Vol. 1. Dordrecht, Boston, London: Kluwer: 195–214.

Malbert, B. 1998. *Urban Planning Participation: Linking Practice and Theory*. Dissertation, Department of Urban Planning and Design, School of Architecture, Chalmers University of Technology. Report SACTH 1998:1. Gothenburg.

Malbrán, H. 1998. *Desarollo Económico, Distribución Urbana y Motorización: Desafíos para el Sistema de Transporte Urbano de Santiago*. Santiago: SECTRA.

Mitchell, G. 2000. Indicators as Tools to Guide Progress on the Sustainable Development Pathway. In R.J. Lawrence (Ed.) *Sustaining Human Settlement – A Challenge for the New Millenium*. London: Urban International Press.

Molebatsi, C.O. 1996. Towards a Sustainable City: Gaborone, Botswana. *Ambio* Vol. 25, No. 2. Stockholm: Royal Swedish Academy of Sciences: 126–133.

Molina, L.T. and M.J. Molina (Eds.) 2002. *Air Quality in the Mexico Megacity. An Integrated Assessment*. Alliance for Global Sustainability Bookseries Science and Technology: Tools for Sustainable Development, Vol. 2. Dordrecht, Boston, London: Kluwer.

Newman, P. and J. Kenworthy 1999. *Sustainability and Cities Overcoming Automobile Dependence*. Washington, D.C.: Island Press.

Nicod, C. and M. Iizuka 2000. *Conciencia ciudadana y contaminación atmosférica: Estado de Situación en el Area Metropolitana de Santiago de Chile*. Santiago: CEPAL.

Perlman, J., E. Hopkins, and Å. Jonsson 1998. Urban Solutions at the Poverty/Environment Intersection. *The Mega-Cities Project, Publication MCP-018*. Hartford.

Presidential Task Group for a Long Term Vision for Botswana 1997. *A Framework for a Long Term Vision for Botswana*. Gaborone.

Rimmer, P.J. 1998. Transport and Telecommunications among World cities. In F.-C. Lo and Y.-M. Yeung (Eds.) *Globalization and the World of Large Cities*. Tokyo, New York, Paris: UNU Press: 433–470.

Rodriguez, A., L. Winchester, and B. Richards 2000. *Santiago*. Working Paper 14 Urban Governance, Partnership and Poverty. Birmingham: International Development Department, School of Public Policy, The University of Birmingham.

Roo, G. de and D. Miller (Eds.) 2000. *Compact Cities and Sustainable Urban Development. A Critical Assessment of Policies and Plans from an International Perspective*. Aldershot: Ashgate.

Rydin, Y., N. Holman, V. Hands, and F. Sommer 2003. Incorporating Sustainable Development Concerns into an Urban Regeneration Project: How Politics can Defeat Procedures. *Journal of Environmental Planning and Management*, Vol. 46, No. 4: 545–561.

Sandercock, L. 2002. Practicing Utopia: Sustaining Cities. *DISP* 148: 4–9.

Sassen, S. 1998. The Impact of the New Technologies and Globalization on Cities. In F.-C. Lo and Y.-M. Yeung (Eds.) *Globalization and the World of Large Cities*. Tokyo, New York, Paris. UNU Press: 391–409.

Satterthwaite, D. 1996. Towards Healthy Cities. *People and the Planet* Vol. 5, No. 2.

Soto, H. De 2000. *The Mystery of Capital: Why Capitalism Triumphs in the West and Fails Everywhere Else*. New York: Basic Books.

Sustainable Cities 1992. Sustainable Cities: Meeting Needs, Reducing Resource Use and Recycling, Re-Use and Reclamation. *Environment and Urbanization*. Vol. 4, No. 2.

The Regional Environmental Center 2004. *What is a Sustainable city? Defined by FARN, Argentina*. http://www.rec.org/REC/Programs/SustainableCities/FarnArgentina.html, retrieved on March 5, 2004.

UN-HABITAT 2002. *Best Practices Database*. http://www.bestpractices.org, retrieved on October 10, 2003.

Weizsäcker, E.U., H. Lovins, and A. Lovins 1997. *Factor Four: Doubling Wealth, Halving Resource Use*. London: Earthscan.

ABOUT THE AUTHORS

Dr. Eran Ben-Joseph is a Professor of Landscape Architecture and Planning at the Massachusetts Institute of Technology. Research and teaching interests include urban and physical design, standards and regulations, site planning technologies and urban simulation. Ben-Joseph has published numerous articles, monographs, book chapters and coauthored the book *Streets and the Shaping of Towns and Cities* published by Island Press in 2003 and *Regulating Place: Standards and the Shaping of Urban America* published by Rutledge in 2004. He has worked as landscape architect and urban planner in Europe, Asia, the Middle East and the United States on projects including new towns and residential developments, streetscapes, stream restorations, parks and recreation planning. He is the founding principal of BNBJ a planning firm in Tel-Aviv, Israel, and E. Ben-Joseph Consultants of Acton, Massachusetts. He previously taught at Virginia Polytechnic Institute and has led national and international multi-disciplinary projects in Singapore, Barcelona, and Washington DC among other places. Dr. Eran Ben-Joseph is the recent recipient of the Wade Award for his work on *Representation of Places: Urban Simulation and the Luminous Planning Table,* a collaboration project with the MIT Media Lab. He holds degrees from the University of California at Berkeley and Chiba National University of Japan.

Pål Castell, born 1974, is a doctoral candidate at the Department of Built Environment and Sustainable Development, School of Architecture, Chalmers University of Technology in Sweden. The main interest of his studies is how civic participation in the management of open public spaces in stigmatized residential areas can influence life quality and sustainable development in those neighborhoods. After completing his master's degree in Landscape Architecture at the Swedish University of Agricultural

Sciences, Pål has been working with projects on Sustainable Urban Design. He has also been actively involved in different organisations of environmental issues for a long time and is currently chair of Chalmers Students for Sustainability and board member of the Centre of Environment and Sustainability at Chalmers and Göteborg Universities.

Dr. Branko I. Cavric received BSc, MSc and PhD degrees in urban and regional planning, urban geography and GIS/LIS from the University of Belgrade, Yugoslavia. His professional, academic and research interests include urban and environmental planning and management, neighborhood planning and design, regional and land use planning, rural planning and GIS. He was a deputy director of the Yugoslav Institute of Town Planning and Housing as well as an assistant professor at the University of Belgrade (Faculty of Geography and Faculty of Architecture) before becoming a member of the Botswana Institute of Town Planners and Urban Land Institute (Washington DC). Since 1993, he has been working as a Senior Urban Planner in the Ministry of Local Government, Lands and Housing of the Republic of Botswana. In 1997, he joined the University of Botswana where he is currently senior lecturer in urban and regional planning and design in the Department of Environmental Science.

Oscar Figueroa, born 1948, is Economist by the University of Chile and has a Doctor degree in Urban Planning by the University of Paris. He is the under director of the Institute of Urban and Territorial Studies of the Catholic University of Chile, where he teaches at its Magister Programs. His main fields of work are urban economics and transportation. He has worked in several Latin American universities, as full or invited professor in Argentina, Ecuador, Brazil, Venezuela, Costa Rica, Guatemala, among others.

Whitney Foutz studied architecture at the University of Virginia before starting her first job in construction management in Washington, DC. She is currently pursuing two master's degrees in Urban Planning and Real Estate Development at the Massachusetts Institute of Technology. Her research interests include emerging real estate markets in developing countries, the role of technology in shaping cities, sustainable urban development and entrepreneurial cities. She has also worked as a research assistant for the Fannie Mae Foundation in Washington, DC and for Community Builders, an affordable housing developer in Boston, Massachusetts.

Prof. Hans J. A. van Ginkel, born in 1940, completed his M.Sc. studies – cum laude – at Utrecht University in human and physical geography, anthropology and history in 1966 and defended his Ph.D. thesis in 1979 – also cum laude. From 1965 to 1968, he was a teacher of geography and history at the Thomas à Kempis College, Arnhem, and from 1968 to 1985, he worked at Utrecht University in the Faculty of Geographical Sciences, where he was appointed full professor in human geography and planning in 1980. From 1981 to 1985, he was Dean of Faculty, becoming a member of the Executive Board of the University in 1985 and in 1986, Rector Magnificus. Hans van Ginkel has a strong interest in the application of geographical knowledge in society, in particular in urban and regional development, public housing and housing markets and public administration. Having also a strong interest in internationalization, he has contributed to numerous international organizations and extended the cooperative links of Utrecht University towards Latin America (in particular, Costa Rica and Bolivia), Southern Africa and Southeast Asia and is a member of the Governing Council of the United Nations University and of UNESCO's Advisory Group for Higher Education. Hans van Ginkel was appointed Rector of the United Nations University in Tokyo in 1997. Since November 1997, he is also a member and Vice-Chair of the Board of Trustees of the Asian Institute of Technology, AIT, Bangkok. He is an active author in his fields of interest and has received numerous medals and distinctions in the Netherlands and worldwide.

Dr. Daniel Irurah, born in 1958, is a Senior Lecturer at the School of Architecture and Planning, University of the Witwatersrand. His focus in teaching, research and consultancy work is sustainability in the built environment. He holds a PhD in Architecture from the University of Pretoria, Master of Architecture and Master of Urban Planning from the University of Oregon as well as Bachelor of Architecture from the University of Nairobi. His PhD thesis focused on the application of the input-output model in analyzing embodied energy of construction materials in South Africa. His master's thesis focused on climatic response, energy conservation and renewable energy for buildings within intertropical regions. His research and consultant work focuses on both environmental and socioeconomic sustainability at various levels of the built environment, including cities, construction, architecture and housing. He also carries out sustainability research on industries allied to construction, especially through the extraction, processing and distribution of inputs (materials and services) to construction.

Dr. Jaan-Henrik Kain, born 1960, is an architect planner with a particular interest in the mechanisms of sustainable urban development. His primary areas of research are complex planning processes, the facilitation of such processes and, more specifically, transdisciplinary management of multifaceted knowledge related to urban development processes. His master's studies in Architecture (1989) focused on processes of self-design and self-construction in rural settlements in Argentina. After receiving his master's degree, he worked as architect and principal of an architecture office. In 1997, he commenced postgraduate studies and received the doctoral degree in Architecture and Urban Planning in 2003. Since 2003, Kain is working as a Chalmers Environmental Initiative Postdoctoral Fellow at the Department of Built Environment and Sustainable Development of Chalmers University of Technology, Gotenborg, Sweden.

Dr. Aly Karam teaches in the planning program in the School of Architecture and Planning at the University of the Witwatersrand in South Africa. He is also the Postgraduate Coordinator for the School. He lectures in the areas of housing, demography, research methods and planning techniques. As a planner, he had the opportunity to see the application of his theories on many projects around the United States, Canada and South Africa and was also constantly engaged with local groups in urban design and environmental issues. Dr. Karam graduated as an architect from the University of Alexandria, pursued his Master's degree in Architecture at the University of Kansas (1986) and his Ph.D. from the University of North Carolina at Chapel Hill in the field of Planning (1993).

Dr. Marco Keiner, born 1963, is the head of the Sustainability Research Group of the Institute for Spatial and Landscape Planning (IRL) at the Swiss Federal Institute of Technology (ETH) in Zurich. He is a senior researcher and lecturer in the postgraduate study program on spatial planning. His areas of work are sustainable spatial development and planning methodology (controlling, benchmarking). Dr. Keiner is also an expert for rural development in developing countries. After finishing his master's studies in Geography (1985–89) and his postgraduate studies in Spatial Planning (1989–91), he received his doctoral degree in Geography in 1999. Keiner worked in cantonal planning offices of Zurich and St. Gall from 1992 to 1994 and then as development expert in Western Africa (1994–98). From 1998 to 2000, he ran his own office and worked for an Austrian architecture bureau in Ethiopia. Since 2000, Dr. Keiner is working with the ETH Zurich.

Dr. Björn Malbert, born 1947, is professor in Design for Sustainable Urban Development within the Chalmers Environmental Initiative at the Department of Built Environment and Sustainable Development of the Chalmers University of Technology in Gotenborg, Sweden. His background is architecture and urban planning and several years of practice as town planning consultant and investigator. His dissertation was on "Urban Planning Participation: Linking Practice and Theory" (1998), which discussed the roles and tasks of professional planners. Currently, he is involved in Urban Development with a focus on urban (infra)structures and the processes of design, knowledge development and decision support for sustainable development from a sociotechnical perspective on urban systems. Björn is project leader for a long-term national research program on Sustainable Urban Development (1997–2004) and Chalmers team leader in EU 5th Framework and AGS research projects. Among others, Björn is a member of the Chalmers Faculty Senate and the Board of the Swedish Research Council for Environment, Agricultural Sciences and Spatial Planning.

Prof. A.C. Mosha, born in 1944, is the head of the Department of Architecture and Town Planning at the University of Botswana. His areas of research interest are urban management, urban poverty, municipal finance, urban agriculture and environmental planning and management, and he has published widely on these topics. After completing his first degree in town planning at the University of Newcastle upon Tyne (UK) in 1972, he went on to pursue postgraduate studies and earned a PhD in town planning from the University of Nottingham (UK) in 1976. He has worked for the Capital Development Authority, Dodoma, Tanzania (1976–78), University College of Lands and Architectural Studies, Tanzania (1978–1990) and the University of Botswana (1990 to present).

Iván Poduje Capdeville, Architect (Universidad Católica de Valparaíso), Magister in Urban Development (Pontificia Universidad Católica de Chile). He is specialized in urban design projects and in the elaboration of strategic plans. In this field, he participated in consultancy project teams for several real estate companies, ministries and municipalities, among others for the new metropolitan zoo of Santiago, the ENEA town, "Costanera Norte," or the rapid suburban railway Santiago–Valparaíso. During three years, he has worked as an associated in the English Spanish planning bureau MECSA, developing urban rehabilitation projects for Santiago, Talca and Punta Arenas. He is a professor at the Faculty of Architecture, Design and Urban Studies at the Pontificia Universidad Católica de Chile, where he leads the magister program on urban development. At the Institute of Urban Studies, he participates in the research network on big cities by publishing and issuing seminar works carried out in Buenos Aires, San Salvador and São Paulo.

Diego Salmerón is researcher at the Division of Landscape and Environmental Planning (LEP) of the Institute for Spatial and Landscape Planning (IRL) at the Swiss Federal Institute of Technology (ETH) in Zurich. His research activities focus mainly on sustainable regional and urban development, including environmental planning issues. He has been working for several years as a project manager in a Training and Research Project in Kunming (China), advising the municipal government and planning authorities on regional and environmental planning matters with regard to the future urban development of the Greater Kunming Area. Salmerón graduated in Rural Engineering at the ETH Zurich. He is member of the Sino-Swiss Society and an associate expert of the Europe-China Management Improvement Foundation (ECMI).

Dr. Willy A. Schmid is Professor for Rural Engineering and Planning, Member of the Directorate of the Institute for National, Regional and Local Planning and member of the Directorate of the Institute for Land Improvement and Water Management in Switzerland. As Vice-Rector for Continuing Education, he is responsible for all postgraduate courses at the Swiss Federal Institute of Technology. Prof. Schmid has extensive experience in working abroad and has published widely in related fields. His key interests are ecological and environmental planning, decision-making in the context of infrastructure and spatial development, development of rural and urban areas and development of geographic information systems as planning instruments.

P. Christopher Zegras has nearly 15 years of experience working on urban transportation, urban development, and environmental issues, with a particular focus on developing world cities. He has worked for MIT's Laboratory for Energy and the Environment and for the International Institute for Energy Conservation, including three years in Santiago de Chile. He has consulted for the Government of Perú, the World Bank, the World Business Council for Sustainable Development and the German and Canadian overseas development agencies, U.S. AID and EPA, among other institutions. He has written or co-written eight peer-reviewed journal articles, several book chapters and numerous reports and working papers. At MIT, he has taught two graduate-level courses, focusing on urban transportation and land planning and development in Latin America. Zegras holds a BA in Economics and Spanish from Tufts University, and a Master's in City Planning and Master's of Science in Transportation, both from MIT. He is currently a PhD candidate in Urban Studies and Planning at MIT.